COVERED BRIDGES
OF THE NORTHEAST

RICHARD SANDERS ALLEN

Dover Publications, Inc.
Mineola, New York

Bibliographical Note

This Dover edition, first published in 2004, is an unabridged republication of the second revised edition (1983) of the work originally published in 1957 by The Stephen Greene Press, Brattleboro, VT. A new Preface has been specially prepared by the author for this reprint.

Library of Congress Cataloging-in-Publication Data

Allen, Richard Sanders, 1917–
Covered bridges of the Northeast / Richard Sanders Allen.
p. cm.
Reprint. Originally published: Brattleboro, VT : Stephen Greene Press, 1957.
Includes bibliographical references and index.
ISBN 0-486-43662-4 (pbk.)
1. Covered bridges—Northeastern States. I. Title.

TG23.1.A44 2004
624.2'18'0974—dc22

2004052891

Manufactured in the United States of America
Dover Publications, Inc., 31 East 2nd Street, Mineola, N.Y. 11501

PREFACE TO THE DOVER EDITION

When this book was first researched, written and published in 1957, a covered bridge—a roofed wooden span over some cascading, rocky brook—was considered a rarity, an ancient relic of the horse-and-buggy days. "Oh, look, John! A covered bridge! Let's take a picture!" Pictures were duly taken, and the bridges sought out by an ever-increasing host of travelers and back road addicts. To them, the spans were stalwart landmarks, with links to a half-remembered past, reminiscent of travel, nature and the crafts that built America. Interest grew, and as the bridges received attention and their features were more closely examined, students of engineering history joined the ranks of nostalgic old-timers, avid antiquarians, and just plain "bridge buffs."

A new era of awareness arrived, just when the general consensus appeared to be: "Covered bridges will soon be gone! Vanished from the landscape! Hurry! See 'em now!" The interest and awareness, mild at first, but further inspired by this book, have burgeoned, and hundreds of Americans now consider covered bridges to be their hobby. They visit them—individually and in groups—photograph them, study them, and try to learn as much as they can about them. Another outgrowth of this renewed interest has been a notable preservation movement. Not only have old covered bridges been saved from destruction and rebuilt, but replicas now occupy dozens of sites. These picturesque architectural American icons, once thought doomed by obsolescence, are now seen as tourist meccas and community assets. Cities, states, whole regions and federal agencies are preserving covered bridges, and, with both private and public funding, building new ones. In fact, at the rate these structures are now being erected, America will soon have as many covered bridges as existed thirty years ago!

The total number of covered bridges in the Northeast today can roughly be set at 185, depending on the criteria of the enumerator. The figure fluctuates with the vagaries of politics, contractors, builders and preservationists. Vermont is said to have the most with ninety-three, while New Hampshire comes in second with forty-nine. Unknown to most New Yorkers, twenty-two covered bridges are to be found in the Empire State. Maine has nine, Massachusetts eight, Connecticut three, and New Jersey just one.

This Dover edition reflects the dawn of awareness and appreciation bestowed on covered wooden bridges, which continue to be logical, practical and economical structures. Presented in these pages is both basic and essential information on covered bridges, together with their historic and engineering background.

—R.S.A.
April 2004

CONTENTS

I

Their Place in Our Past

There is something about covered bridges. To a relatively small handful of zealots they represent the artistry of our forefathers and so must be preserved at all costs. To thousands of tourists they are rarities well worth braving horse-and-buggy roads to find. To millions of Americans who have never seen one they have become the symbols of the quiet and simplicity of a happier day.

It is not hard to define their initial charm: it is because they are roofed. Yet it would fail to do justice to covered bridges if it were assumed that the roofs were added as an architectural flourish.

Why were bridges covered?

The question seems to be as old as the bridges themselves. Well, they were not covered to protect the user, his horse or his load of hay. They were not covered so that horses would think they were barns, or to prevent old Dobbin from shying at the glint of water. They were not roofed just because their builders were trained in putting up barn-like structures, or because bridges with roofs were preferred over iron bridges which might attract lightning. Nor were they covered to keep snow off the floor: old town reports dispute this idea with their entries for "snowing the bridges." And certainly they were not covered, as one tongue-in-cheek theory has it, to prevent a traveler's knowing what kind of town he was approaching until it was too late to turn back.

The real and only reason for covering bridges—discounting all tales, theories and legends—was to protect the wooden skeleton and thus preserve the bridge itself.

Early builders, who were experts in wood, understood that its great enemy is moisture. Wood that is alternately wet and dry soon rots. The bridges were covered to keep the inside structural timbers dry. If they were kept dry the bridges usually lasted. Floors were expendable: the big things to protect were the supporting sides, or trusses, which gave the structure its strength.

That's all there was to it. Ask any old-time New England carpenter why they covered bridges and he'll tell you:

"Why did our grandmothers wear petticoats? To protect their underpinning. Why did they cover bridges? Likewise."

But even such a sound informant may be hazy about where you could find a worthwhile specimen, unless one still existed in his own bailiwick. Like most people, he'd allow that covered bridges are seen here and there; but also like most people he might be under the impression that they are nearly vanished landmarks. This is certainly not the case. Recent surveys show the existence of well over one thousand covered bridges in the United States and Canada. In some areas of southern Pennslyvania and Ohio, and in rural Quebec they are a not uncommon means of crossing the many streams.

Rock-ribbed Yankees will probably be surprised to learn that 80 percent of all covered bridges standing today are located outside their territory. Pennsylvania leads all her sister states without a struggle for supremacy, having over 200 covered spans within her borders. The nearest competitor is Ohio with well over 100, followed by Indiana with just under 100.

The number of covered bridges in the Northeast has dwindled to 192. This figure includes 7 railroad bridges and 185 highway spans—ones big enough to carry vehicular traffic—of which the great majority are on public roads, several are open to the public as historical exhibits and a handful (also full-sized) are on private property. Vermont has the most, with 92 highway and 2 railroad bridges. New Hampshire is second with 46 plus 5 carrying railroads. Unknown to most New Yorkers, 25 covered bridges are still to be found in the Empire State. Massachusetts has 9; Maine, 9; Connecticut, 3, and New Jersey just 1. Rhode Island, where some of the greatest of these bridges once stood, has none left at all. An inheritance from early America, these existing Northeastern spans represent covered bridge types everywhere and are, in many instances, examples of great advances in bridge engineering.

THE THRILL THAT COMES ONCE IN A LIFETIME

More people than you'd suspect can recall the covered bridges they knew and delighted in during their youth. Children, especially, have always been attracted to covered bridges. The smell in the darkened tunnel was a delicate aroma of wood shavings, ammonia, hay and horse manure—hardly Chanel No. 5, but a scent that, once sniffed, could never be forgotten. After a trudge in the hot sun, bare toes sifted the cool dust on the bridge floor and knew a wonderful feeling. Light, reflected from the water below, would flicker up through the floorboards to make dancing spots on the rafters, shimmering and ever-changing. A worm-baited fishline dangling down through a hole in the floor sometimes produced a whopper.

The rafters were a great place for acrobatics, and to be able to swing like Tarzan hand over hand from beam to beam was an accomplishment only the biggest boys could master. The more daring of the girls, especially the tomboys, would compromise their dignity by skittering up and over the wide flat arches, skirts flying.

Experience determined whether you loved a bridge or held it a bit in fear. To cross at night was like passing a graveyard. You could always imagine that there were robbers hiding high above you in the dark, just waiting to drop down upon the unsuspecting traveler. A good hearty whistle often served to bolster a small boy's courage as he tramped as fast as his short legs could carry him through the dark shadows of a long bridge, for to run would be to admit you were scared.

Snug at home in bed you could listen to the rumblings and hoofbeats, and learn to tell who was crossing. Clop! Clop! Clop! That would be Deacon Haverly on his way home from prayer meeting. A rattle and flying hoofs might mean the doctor was racing the stork. Clop! Clop! Clump! Clop!—George Kepler and his mare with the spring halt. A very slow walk. Ummmmm. Must be the storekeeper taking the Faraday girl home. Finally a rumble and roar, pop and bang marked the passage of Charlie Oakes' new Packard with the carbide headlamps.

A rainy day and little traffic brought a game of "one-o'-cat" to the bridge. If you slammed the ball beyond the sheltering portals you were out. When the game grew tiring there were always hundreds of initials to be examined and chalked notes of import to their inscribers to be deciphered: "S.E.C. 1880—W.C. 1885—BL—FM 1900—F. Brown, August 1892—I Hugged Polly P. in this Bridge—Liar—Did too—Didn't Didn't *Didn't.* F. B's a Liar! —Did too."

There was a lot of history in these hieroglyphics, and mysteries you could penetrate if you knew the townspeople and their ancestors.

Best of all for rainy day perusal were the old advertisements. The roaming advertising man of the past century used to hitch up his mare and fill his wagon full of placards that named his product and extolled its virtues. Then he set out, stopping to nail his notices on trees, fences, barns, sheds, outbuildings and covered bridges. The gaudy signs gleamed in the Spring sunshine, faded in the August glare, began to rot and flap loose in the Fall rains and often vanished entirely in the blasts of December.

Yet inside the covered bridges they stayed bright, tacked securely, dry and unweathered as the years went by. Only as the bridges themselves aged did the advertisements mellow and fade. Within the protecting confines of the covered spans were found ads for the forgotten patent medicines, nostrums, foods, tobacco, cooking utensils and various personal services of the '70's and '80's. There was Kickapoo Indian Oil, Friend's Baking Powder, Cinco Cigars, Rogers Stoves, Battle-Axe Plug

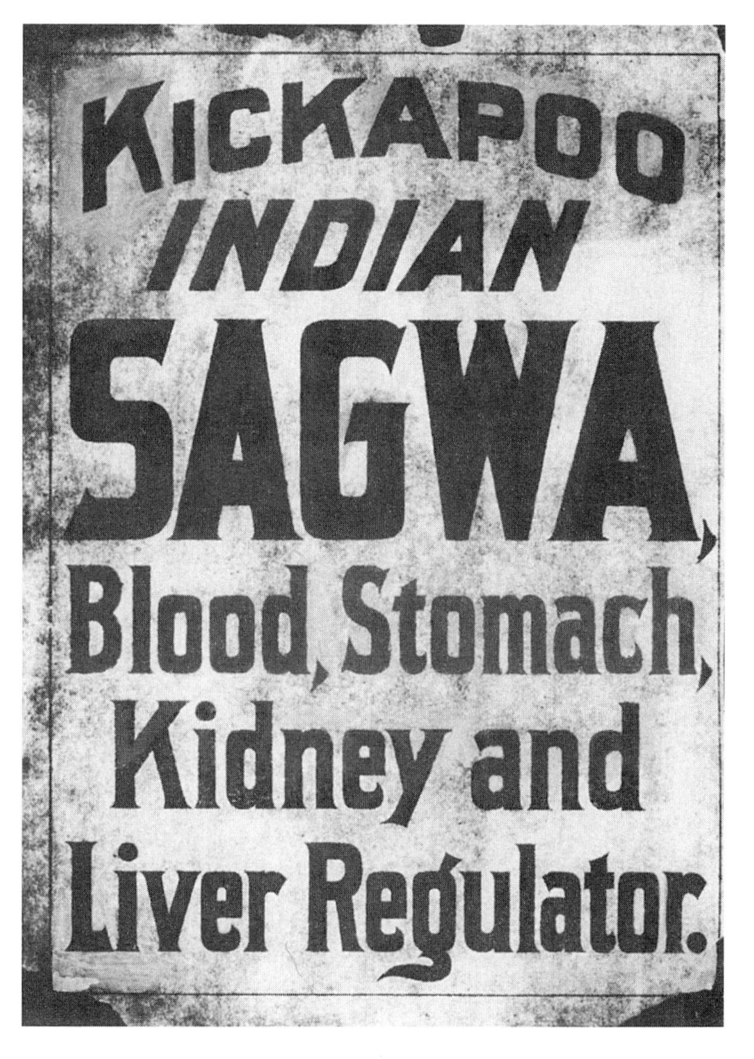

Tobacco—a list of them all would fill a book by themselves.

Particularly notable and well-advertised throughout the Northeast were Kendall's Spavin Cure and Dr. Flint's Powders, both beneficial to man's trusty friend the horse. These two products carried the fame of tiny Enosburg Falls, Vermont, to far corners of the countryside.

If the bridge was on a main road leading to town, a local merchant was wont to place a large, well-painted sign on the portal, calling attention to his wares. Another method was to use the cross timbers inside the bridge

for a progression of signs in the manner later made famous by Burma-Shave. In those days the occupants of a buggy, jolting through the long tunnel, turned their eyes upward to read on successive beams:

"Whitehill's . . . Cor. Main & River . . . Ladies & Men's . . . Clothing . . . Shoes, Too!"

The county fair, an annual event, had to be careful about its ads in the bridges. One old-time farm family went to the fair a week too late as a result of heeding a year-old poster whose message remained still bright and inviting inside a bridge they seldom crossed.

As for color, the circus reached the zenith in covered-bridge advertising. The advance man for the Big Top usually chose the weather-boarded wall spaces just inside the entrances for his full-sized posters. Sometimes most of the bill was a gigantic numeral of the date the circus was to show; but more often lions and tigers, hippopotami, cavorting clowns or some beautiful damsel on the flying trapeze—clad in pink tights—regaled the youngsters who played in the bridge, offering enchantments in no way diminished as the posters peeled and faded with the passing years.

Then there were broadsides plastered close to eye-level along the sidewalk railing to entice country girls to seek the "ample wages and short hours" of the Big City's shirtwaist factory. Perhaps the acme in covered bridge advertising was reached by members of a religious group many years ago. The bridge on which they placed their sign stood at a right angle to the foot of a long hill. Over the portal was emblazoned in bold letters: Turn or Burn!

But most warning signs read something like this: Five Dollars Fine for Riding or Driving on This Bridge Faster Than a Walk. Although it sounds funny today, this admonition had a mighty good reason behind it. The concentrated beat of horses' hoofs, coming down simultaneously, did far more injury to a covered bridge than overloading it beyond its capacity. Fear of damage to joints from such a shaking is the reason why soldiers were always ordered to break step while crossing a bridge, and the same injunction applied to horses.

Covered bridges have always had a romantic appeal to both old married couples and young people on their first dates. From time immemorial they have been the darkened tunnels of romance. Ask grandpa why they called them "kissing bridges" and you'll be rewarded with a reminiscent chuckle. Even though the structures were open at each end they gave a certain privacy. Couples on foot or in a buggy could easily squeeze hands or exchange a kiss in the darkness.

Of course there was always the possibility of unknown observers. One boy found it in-

teresting to secrete himself high in the cobwebbed rafters and eavesdrop on those who took an afternoon promenade through the bridge. One Sunday he was astonished when his sister Mehitabel, in company with the young minister, stopped right below him.

"Hitty . . . ," stammered the flustered dominie.

"Yes?"

"Hitty . . . well . . . Marry, will you Hitty me?"

The boy in the rafters couldn't contain himself and burst out with a half-smothered snicker. Startled red faces stared up at him, and the minister gave rein to the horse. Armed now with secret information, the boy carried on a mild campaign of blackmail against his sister—but his supply of spruce gum and Jackson Balls was cut off abruptly when Mehitabel up and married the preacher.

The covered spans served as shelter for all kinds of meetings, from political rallies for Tilden or for Hayes to bond rallies during World War II. Church services were sometimes held in the bridges on a hot summer Sunday, and church suppers were spread on long tables laid from one end to another of more than one little-used bridge. Roped off temporarily, an isolated covered structure became a good place for two belligerent young bloods to settle a grudge fight. Still another bridge was used by a woman who lived beside it, for drying clothes on rainy days.

Each covered bridge holds memories, large or small, and added together these recollections amount to a lot of affection for the old spans. When they go, a part of our past goes with them.

II

The Bones of a Bridge

LONGFELLOW described a covered bridge as "a brief darkness leading from light to light." It is this brief darkness that so few people pause to examine, even though they may love these dusky tunnels. For the interior of any bridge tells the traveler just *what* the early builders were protecting with their roofs and siding. By studying its skeleton from the inside you can tell what a bridge is all about.

To many casual observers one covered bridge looks very much like another, yet how much more pleasure these bridges give to those who know and respect their design! Liking a bridge because it makes a delightful picture is a worthwhile sentiment so far as it goes. But no antique collector would think of a fine piece by Sheraton or Chippendale as a quaint little chair, and no true friend of covered bridges likes them only for their artistic effect and ignores the manner of their construction. Even in this atomic age our covered bridges still serve as logical, economical and efficient structures. They deserve to be saluted for their engineering as well as for their charm.

A bridge is defined simply as a structure erected to furnish a roadway over a depression or an obstacle—that is, over valleys or chasms, over water or other roads.

In the dim past of European history, long before there were any stone bridges, there were wooden ones. When he wanted to cross a narrow stream the earliest bridge builder probably chose a good stout tree growing on one bank and then felled it to make it cross the gap to the other bank. In later years a simple log crossing of this type came to be called a *stringer bridge,* and its shore foundations—for

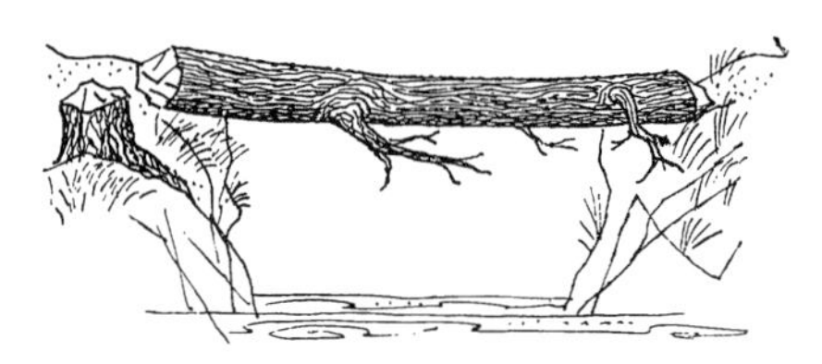

men soon began to improve upon the rough banks provided by Nature—were termed *abutments.*

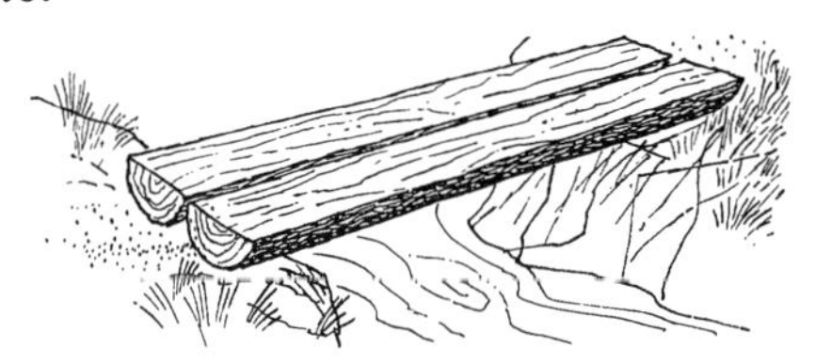

The next step in bridges was the discovery that a log stringer split in half lengthwise, with the open halves laid side by side across the stream, gave the traveler better footing. A further refinement decreed that the stringers be separated and short logs laid across them to form a wider walkway.

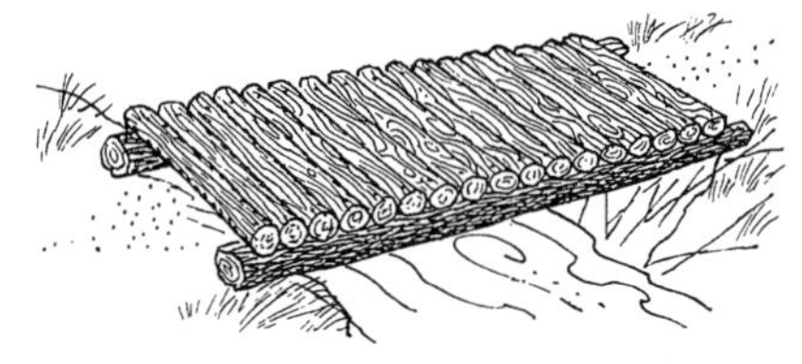

So here was one basic plan for a bridge. It is a plan used successfully in modern short bridges whose stringers, now steel girders or prestressed concrete slabs, rest on the abutments of either bank and receive no other support.

But what if a stream was wide? Of course, longer logs could be placed across it, but the longer these stringers had to be, the more likely they were to sag. It was in the remote past when unknown geniuses of Central Europe evolved a means to counteract the sag.

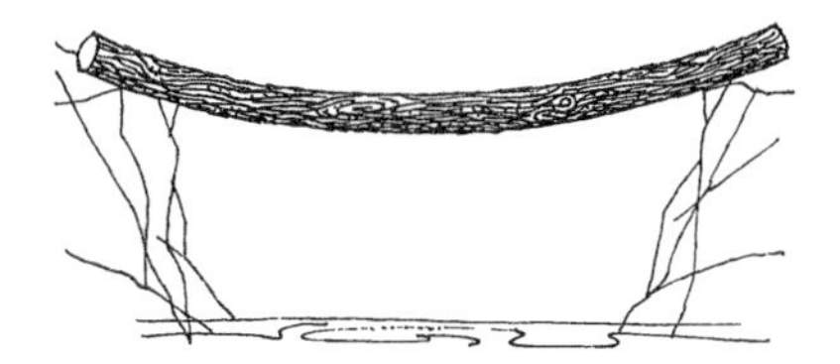

First, each stringer was supported from underneath with two *braces,* diagonal supports slanting toward the midpoint of the bridge. To do this builders cut two logs, pressed their butts into either bank and sloped the tips upward so that they met at an obtuse angle in the center of the stringer in the form of a Λ. Next, a crosspiece—in the form of a parallel stringer set lower against both banks—was added to close the open end of the Λ and prevent the diagonal arms from shifting. This horizontal piece came to be called

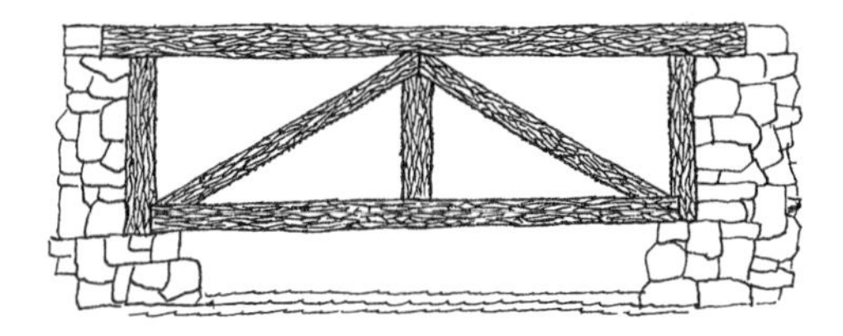

the *lower chord;* the original stringer was termed the *upper chord.* This, then, was a *truss,* a triangular system of timbers so devised that each timber helped support another, and together they supported whatever weight was put upon the whole. Later still a center post made its appearance, reaching from the apex of the Λ down to the midpoint of its new base. This simple arrangement of timbers was given an obvious name—the *kingpost* truss. It is the earliest formal bridge truss design. Although they may not have realized it, these ancient builders were employing a primary engineering principle:

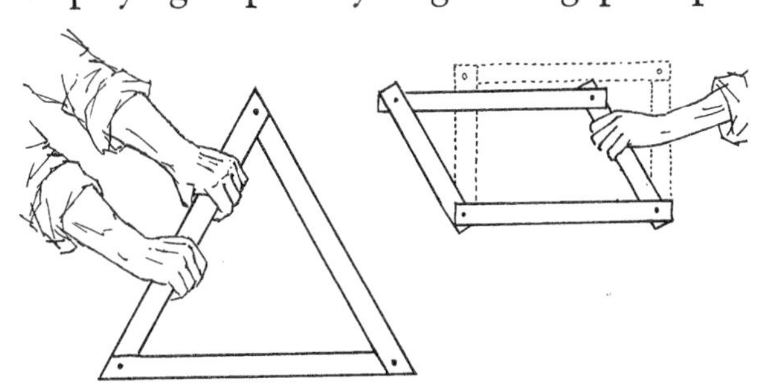

under a load a triangle will hold its shape until its side members or its joints are crushed. A frame of more than three sides will shift out of shape under pressure unless it is made rigid by internal braces, in which case its rigidity derives from its being turned into an arrangement of triangles.

In that primitive kingpost truss we see that the bridge was braced by one big triangle, and that, with the introduction of the center post, the supporting framework actually amounted to two smaller right triangles back to back.

The earliest kingpost trusses were built *under* a stringer bridge. This made them vulnerable to ice and high water. Erection was difficult, and even impossible, at many sites. We can't pinpoint the time, the place or the builder who first realized that the principle of the kingpost triangle applied just as readily to a truss built *above* the stringers.

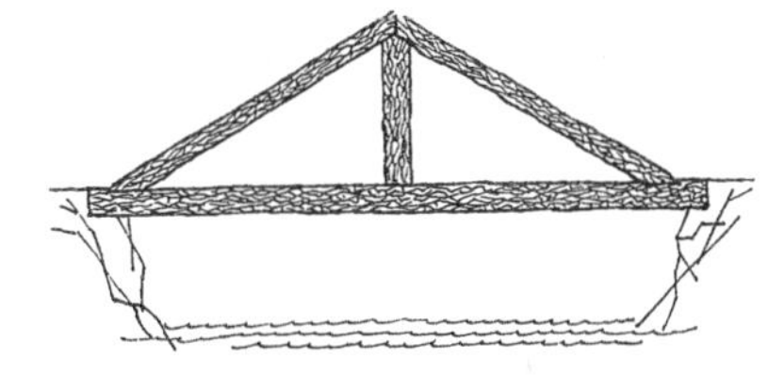

Our first primitive little kingpost bridge of Central Europe used the truss as a *sub*structure, and such bridges came to be designated as *deck trusses.* These were never developed to any great extent in America. Bridges where the truss was a *super*structure were termed *through trusses,* and most of the bridges examined in this book are of the through variety.

In a through truss kingpost bridge the diagonally arranged braces of the Λ are called *compression members,* compression being the squeeze these members are subjected to as a load passes over the bridge. On the other hand, the center or kingpost and the lower chord are called *tension members,* because when a load crosses the bridge they are pulled downward and are subjected to tension. This explanation is really an oversimplification of the stresses that truss members un-

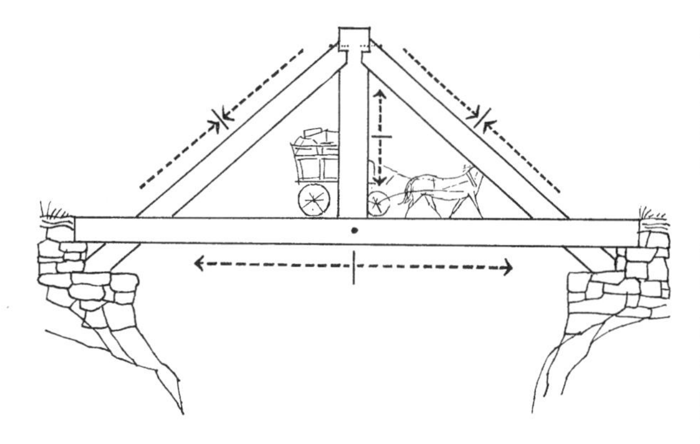

dergo, but I believe it is enough for our purposes now in describing the functions of truss parts. Indeed, it wasn't until the middle of the 19th century that the actions and stresses of members in elaborate trusses were fully described. This whole matter is extremely complex, especially when one realizes that the progress of a load across a bridge creates additional problems of variations in stress.

The triangular form of the kingpost served as a basic design for many uncovered wooden bridges in Europe and America, as well as for a number of covered ones. Any carpenter worth his salt could—and did—erect a kingpost truss bridge across a stream in a hurry; after all, he was accustomed to using the familiar triangle in the roofs of houses and barns.

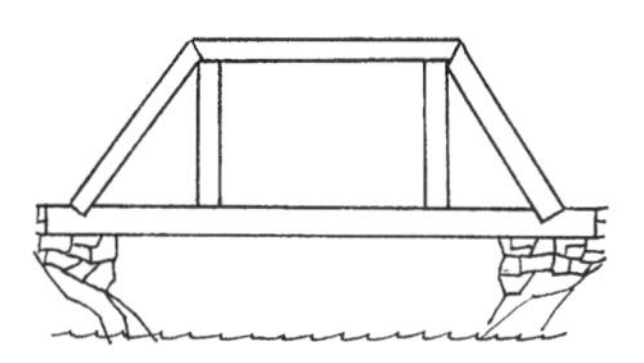

A natural development of the kingpost was the *queenpost* truss, the better half of a royal family. If you demonstrate a kingpost by placing your index fingers together in a peak and joining your thumbs to form a base, you can just as easily see what a queenpost is by placing a match or pencil stub horizontally between the tips of your index fingers and joining your thumbs as before. In a queenpost you have replaced the peak of the kingpost with this horizontal crosspiece, and have allowed the base to become longer. The queenpost created a truss capable of spanning wider streams than the kingpost could manage to do.

When it became necessary to throw a single span across wider streams the next logical development was the *multiple kingpost.* As its name implies, this truss was a series of upright wooden posts with all braces inclined from the abutments and leaning toward the center or kingpost. A corollary design was the combination of an arch and a kingpost truss. These more complex trusses allowed several stringers to be spliced together for a much longer bridge. Thus the triangle continued to be used

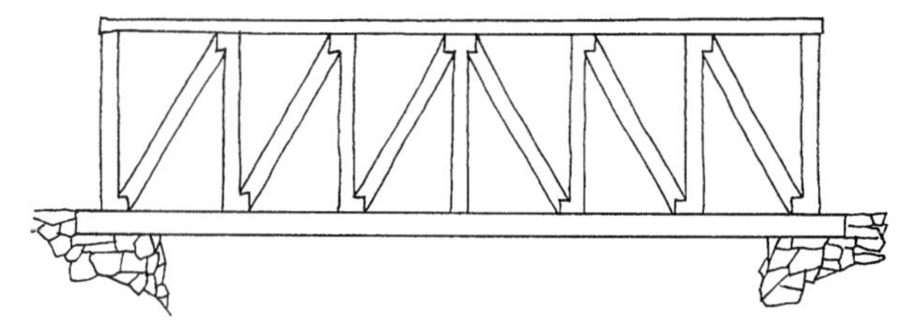

in bridge construction; and it is a basic form still used in our giant spans today. These simple trusses were first described formally by an Italian architect of the post-Renaissance named Palladio.

For three hundred years, from the 16th to the 19th centuries, wooden bridge design lay dormant. Then in pioneer America—with the need to cross our major rivers with bridges of a size and length hitherto undreamed of—less than fifty years saw wooden bridge design brought to its peak.

The first American innovation was to amplify the ancient use of a wooden arch to strengthen a basic multiple kingpost; this was the *Palmer* truss, followed in short order by the basically similar *Burr* truss. But in the 1830's bridge building in our Northeast brought about yet another discovery—i.e., with careful and more intricate trusswork, the arch could be dispensed with entirely in favor of a *panel* truss, at first referred to as the *Long* truss after its original inventor. A Long truss was composed of a series of boxed X's, with

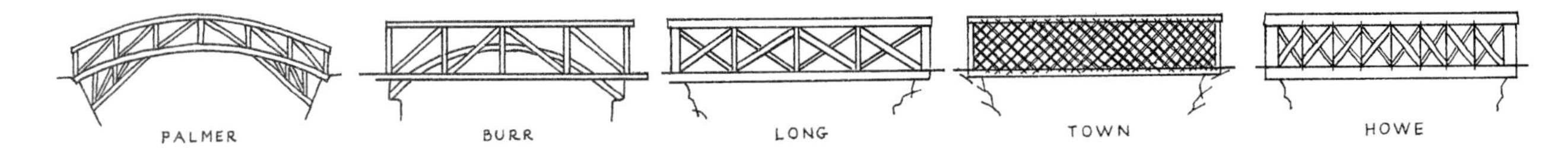

three or more panels comprising the entire truss.

A deviation not considered an outgrowth from any one truss plan was the *Town lattice* truss. This was a series of many overlapping triangles, using no arch and no uprights, which could be built in a continuous network and was able to support spans up to 200 feet in length. There is a paradox in the contention that the Town truss owed nothing to preceding designs: it did discard the uprights of the kingpost and the panel and ignored the arch, but in doing so it demonstrated that the simple triangle is the ideal supporting device. It was this emphasis on the basic, unadorned principle which made the Town truss unique. It was light, easily constructed and incredibly strong despite its fragile appearance.

In 1840, near the end of this half-century of concentrated development, came the *Howe* truss in which adjustable iron tie-rods were substituted for the old wooden uprights of the panel variety, thus marking the transition between purely wooden structures and those built entirely of iron. These American trusses and their impact on bridge construction must wait until the next chapter to be described in detail; at the same time the inventors will be given credit, long overdue, for their masterpieces.

So far the bridges discussed here have had only one *span,* meaning the length of the bridge from abutment to abutment. The *clear span* is the distance between the face of one abutment to the face of another, while the *truss span*—the expression most commonly used—denotes the length of a bridge's truss regardless of how far it may extend over the land beyond the actual abutments.

When a waterway was too wide for one span the additional bridge trusses had to be supported by piers, either the natural outcroppings of rock or man-made ones. These structures were called *multi-span* bridges and could be of the type best suited to the builder's talents—simple stringer, king- or queenpost, arch-truss or lattice—provided that the current was not too swift, the water not too deep or the river bed not too unstable.

With the intricate wooden skeleton to protect from the weather, it is no wonder that village carpenters evolved the idea of preserving the timbers of a bridge by covering the structure with a roof. True, builders in Biblical times roofed their bridges. But ancient artisans added covers mostly to shelter the people who crossed or congregated on the bridges; and in Italy, China and Mexico stone bridges with wooden roofs are used as market-places. It was the men of Switzerland and Germany, those who devised the kingpost and queenpost, who are also recognized as the first to cover their bridges primarily to protect the trusses. Covered truss bridges abounded in the heavily forested regions of Europe as long ago as the 14th century. Several covered bridges are still to be found in Switzerland, built of giant virgin timber beams in king- or queenpost design, and they have been standing more than four hundred years.

It is progress, in the form of speed and modern roads, which has brought neglect or destruction to far too many covered bridges in America. But the imaginative engineering that went into the staunch bones of our covered spans promised that, on some later day, their steel-webbed successors would soar high and far across our mighty waterways.

III

The Builders:

Unsung Pioneers of Engineering

BUILDERS never did come in for much credit. Who knows the name of the man who actually was responsible for building the big business block in the center of town or the railroad line across the state, or even who laid out and constructed the new airport of two decades back? Thus it was with this region's covered bridge builders. People took their work for granted. Elsewhere in the country you will find that covered bridges flaunted the names of their builders, neatly painted in bold black letters on the portals or carefully inscribed on a stone in the abutments for future generations to read. But with few exceptions the builders of the Northeast were content to let their bridges stand in mute testimony to their skills.

We have seen how wooden trusses and wooden roofs for bridges were originally developed in Europe. Was any one builder responsible for the introduction of the covered bridge to America? The mists of antiquity have never cleared enough to give an absolute answer. However, there was one architect of the Italian post-Renaissance period who certainly had a lot of influence on our bridges: Andrea Palladio, who drew the first published plans of wooden bridge trusses to receive wide distribution.

ANDREA PALLADIO

Palladio would pass his thin hand across his large and luminous eyes in disbelief were he to have seen some of the great wooden arch-truss bridges that once stood in America. And he'd probably have dropped his calipers in amazement if he could have been told that they were lineal descendants of a little wooden fabrication he erected near Bassano, Italy, over a mountain torrent in the foothills of the Italian Alps. But how he came to be called "the grandfather of the American covered bridge" can be partly reconstructed.

Andrea Palladio was born in 1518 at Vicenza in Northern Italy. He grew to manhood in a period when promising and talented youths were taken under the wing of some patron, often a nobleman or rich merchant. Young Palladio studied mathematics and did a bit of sculpturing, but soon devoted himself

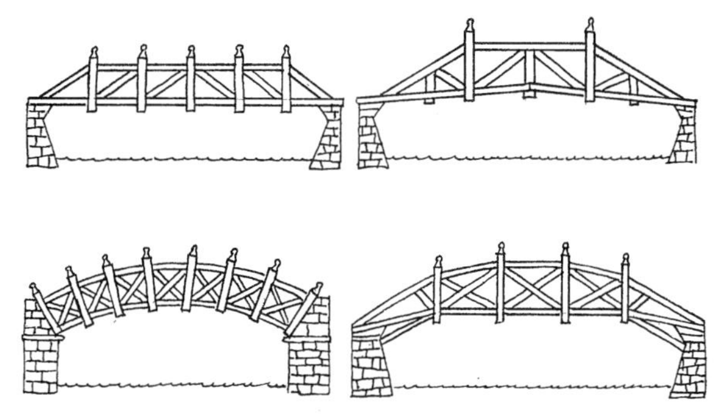

The four basic bridge truss designs described by Palladio in Treatise on Architecture.

to a detailed study of architecture. While he was still a comparatively young man he applied early Greek and Roman designs to buildings of his day, and this style of architecture has since become known as the Palladian school. His wooden bridge plans are contained in a four-volume work, *A Treatise on Architecture,* completed in 1570. In the third volume he shows four different applications of the truss system to bridge building; his are the first known illustrations of such designs. Palladio did not claim to have invented the trusses, and states that a friend, Alessandro Picheroni de Mirandola, had seen bridges of these types in Germany—which is entirely probable. Yet, being a delineator as well as an architect, Palladio took the plans out of the mind's eye and put them on paper for others to follow.

In one design he used a basic queenpost such as was found in Switzerland and Germany. Palladio actually used this plan in building his first bridge, a 108-foot span over the Cismone near Bassano. Another of his designs showed a multiple kingpost, the logical next step toward spanning streams wider than it was possible to cross with the two "royal family" trusses. The architect's other two drawings showed wooden bridges combining an arch and a truss.

European builders paid scant attention to Palladio's timber truss designs, and they were half-forgotten for nearly two hundred years. Finally, in 1742, Palladio's works were translated into English by Giacomo Leoni. At about the same time another book on architecture, by a Englishman named William Gibbs, reflected strongly the influence of the Italian master. A great revival of Palladian architecture swept England, sparked by Lord Burlington, who even had an 87-foot bridge on the Palladian arch-truss design erected in the garden of Wilton Park, his estate. Still, stone was the solid and permanent bridge building material in most of Europe. Sizable, workable wood had become scarce.

In America things were different. Right after the Revolution the United States were united in name only. Great distances separated the states. Coastwise shipping was roundabout, and travel at best was downright inconvenient. The rivers were the obstacles to land communication up and down those thirteen states. Not only did the big streams all drain into the Atlantic, they were broad and long and treacherous. A bit of high water would make a fording place useless in a few hours, and the traveler, coming to a ferry, seemed always to arrive at a time when the old scow had just set out for the other side.

Bridges just *had* to be built, and quickly, if the scattered settlements of the country were going to be held together. Stone bridges were out of the question: they took too long to build and were too expensive. But there was plenty of wood, giant timbers of virgin white pine, the like of which had gone into the masts of His Majesty's ships before the late war. They'd be big enough for some enormous bridges.

The new government was too hard-pressed to undertake the job, so enterprising individuals saw their opportunity to turn an honest dollar. They set about forming toll-bridge companies, chartered to erect substantial bridges across the Merrimack, the Charles, the Connecticut, the Hudson and the Delaware. At first they were content to have long trestle bridges of many spans set on pilings and close to the water. But year after year the proprietors would lose their bridges (and their shirts) in the annual Spring freshets. Piling ice and high water would embrace a trestle bridge and quickly carry it seaward in little pieces. The men who operated them had to pour dollar after dollar back into their investments to break even.

In 1764, two decades after Palladio's designs appeared in an English edition, a "geometry work bridge" had been erected over the Shetucket River at Norwich, Connecticut. The bridge, 124 feet long and 28 feet above the water, was the work of John Bliss, "one of the most curious mechanics of the age." Called Leffingwell's Bridge, it is thought to be the first application of the truss to bridges in America.

Still, what the bridge corporations needed was a new design, one capable of spanning longer distances than those managed by a king- or queenpost, and one with no intermediate supports that would block a channel. Some of the best minds of the time set to work on the problem. There was Thomas Pope, a New York landscape architect who dreamed up his "Flying Pendant Lever" bridge which he confidently proposed to fling across the Hudson River from Manhattan to the Jersey shore. Even Thomas Paine, the political pamphleteer, and Charles Willson Peale, the eminent artist, joined in the free-for-all of bridge designing, plotting out fanciful spans which were never to be built except in their inventors' fertile imaginations.

At last, up in Newburyport, Massachusetts, a local man, Timothy Palmer, got on the right track. It seems pretty evident that he must have come across one of Leoni's English translations of Palladio's *Treatise on Architecture.* Huge squared timbers, lapped and mortised to form wooden arches, were Palmer's answer to the long-span bridge problem, and he became their first successful designer. Palladio had erected his bridges over tiny Alpine torrents, but Timothy Palmer translated the old Italian's brain children into

TIMOTHY PALMER

great timber arches made from the giant stands of pine in New England forests.

Described in some accounts as "a playful eccentric" and in others as "self-taught architect, ingenious housewright and inventor," Timothy Palmer was a big man with a prominent Roman nose and hair tied in the fashionable queue of the time. Apprenticed to the Spofford family of ship- and millwrights, he was over forty before he commenced to build bridges. His first recorded attempt was in 1792, bridging the Merrimack River between the shores of the mainland and Deer Island, three miles above Newburyport. Here Palmer erected two arched spans, one of 160 feet and one of 113 feet, one on each side of the island.

In the absence of any written account by Palmer that he was influenced by Palladio's bridge designs, it cannot be certain that he used the Italian's drawings as a basis for his work. Perhaps there was no copying as such, but only a vague notion as to the shape a proper bridge of long span should take. Nevertheless both Leoni's and Gibbs' works on Palladian architecture were doubtless accessible to Palmer in his apprentice days.

The next year the "ingenious housewright" built Andover Bridge (at the present site of the city of Lawrence, Massachusetts) and then crossed the Merrimack a third time with the three-span bridge at Haverhill. Other bridges built by Palmer and his employer, Moody Spofford, crossed the Connecticut River at Windsor, Vermont, and the Kennebec River at Augusta, Maine. Their biggest achievement in New England was the Great Arch of the Piscataqua seven miles above Portsmouth, New Hampshire. Actually, this bridge spanned the Great Bay *beside* the Piscataqua River, connecting two rocky islands with the shores of Newington and Durham. The whole bridge was 1362 feet long, mostly pile-and-trestle, with an immense 244-foot trussed arch over the deepest channel. The curved timbers of the arch were hewn especially from crooked logs.

Timothy Palmer's bridges were all similar to one of Palladio's arch plans in which the roadway followed the hump of the curve. Where there was more than one span, users had the same sensation as riding over a series of small hills in rapid succession. They were *not* covered bridges to begin with, though the smaller span of the Deer Island Bridge was later housed and roofed. The inventor secured a patent on his arched truss in 1797 and began to build his bridges farther afield—over the Potomac, the Schuylkill and the Delaware. His 550-foot, three-span Permanent Bridge at Philadelphia, finished in 1805, was the first known American covered bridge. Palmer was proud of the intricate big trusses in his designs, and wanted them to show for the examination of prospective builders. But Judge Richard Peters of the bridge company had the idea that enclosing the great wooden timbers from the weather would add greatly to the life of Philadelphia's long-wanted and fine new bridge. The company commissioned Adam Traquair and Owen Biddle to design and build this covering. Timothy Palmer gracefully conceded that, with this newfangled refinement, the bridge perhaps "*might* last thirty or forty years." It actually stood for forty-five until replaced by another covered bridge built to carry a railroad. Thus a Massachusetts architect, a Philadelphia judge and some local carpenters were responsible for America's first covered bridge.

The Newburyport builder must have become a quick convert to weather protection for his work. His next bridge, across the Delaware River between Easton, Pennsylvania, and Phillipsburg, New Jersey, was tightly enclosed and roofed at the time of its completion in 1806. After finishing these "showcase bridges" in Pennsylvania and New Jersey Timothy Palmer returned quietly to Newburyport to finish out his life in 1821. For a "playful eccentric" the old master builder did mighty well.

Palmer's Deer Island Bridge over the Merrimack was built in 1792 and roofed around 1810.

Though Timothy Palmer used an arch similar to the Palladian one in his bridges, it was another New Englander—Theodore Burr—who really extended the use of this type into other parts of the country. Burr was born in 1771 in the little Connecticut hill town of Torringford. His father was a millwright and young Theodore took to heavy construction naturally. After a good schooling and apprenticeship in the building trade, Burr became a pioneer and went "west" to what was then the wilds of New York State and became one of the first settlers of Oxford in the Chenango

Union Bridge at Waterford, N.Y., typified Burr's patent design with its arch treatment and level roadway.

Valley. There Burr built a saw-and-grist mill, and, since customers came from both sides of the Chenango River, it was only natural that his next project should be a bridge. Commencing with his first simple stringer bridge at Oxford in 1800, Theodore Burr was seldom out of sight of a river to be spanned.

An easy-going optimist, Burr was willing to tackle almost any bridge building job, though at first he didn't go too far for his contracts. Gradually his fame grew and he began getting requests from the directors of toll-bridge companies all over New York State. He put up a drawbridge at Catskill, a multi-span structure over the Hudson River at Fort Miller, and a 330-foot arch bridge across the Mohawk at Canajoharie.

Experimenting with different bridge types fascinated Burr. With the erection of the one over the Hudson River at Waterford, New York, he found a design which suited his fancy. This was an adaptation of one of the old Palladian arches, in this case one with a level roadway. Four spans gave this Union Bridge at Waterford a length of nearly 800 feet. Burr thought well enough of this design to have it patented in 1806. Called the *Burr* truss, it was the prototype of thousands of other covered wooden bridges all over the country. In the Burr plan the great arches were made more rigid and sturdy by framing them outside, inside or between with a series of multiple kingpost braces and counterbraces.

Theodore Burr was the kind of man who could sit down with bridge company directors and casually scribble the amount of his bid, involving thousands of dollars, on a piece of scrap paper. While he talked he would cover the edges of the paper with pencil doodles, and then, as if it were an afterthought, he would add a note at the bottom—"Superintendence, say $6,450.00."

Burr built dozens of bridges in New York State and New Jersey, and became known as "the celebrated bridge architect." In 1812 he moved to Pennsylvania, where he added to his fame by building five of his gigantic covered bridges across the wide Susquehanna River—with all five jobs going on simultaneously! It would be fine to report that Theodore Burr made his fortune in bridge building, but it would not be true. Overreaching himself by trying to do five jobs at once, hounded by creditors at every bend of the Susquehanna, the bridge builder lost his capital and was driven to an early death in 1822. There is no record of his burial place. With his family too strapped for funeral expenses, his body may lie in some Pennsylvania potter's field.

But Theodore Burr's patent arches lived on. His boss-carpenters became contractors themselves, bridging the countless lesser streams of New York and carrying the design back into New England, the birthplace of the inventor. Theodore Burr once told a board of bridge company directors: "If you would have a good structure, examine the works of many famous bridge builders and try their talents in building bridges over a river—instead of building on paper." Followers of Burr's plans certainly proved his point.

The big timbers that went into the pioneer bridges of Palmer, Burr and lesser builders took considerable manpower to put together. There was a crying need for a substantial bridge that could be erected by a common carpenter's gang. Ithiel Town, who was born in Thompson, Connecticut, in 1784, and became an architect in New Haven, had the sense to foresee this need and do something about it. Town had already built the first covered bridges across the Connecticut River at Hartford, Springfield and Northampton, using Burr's arch type of construction. He came to the conclusion that there must be an easier way to make a wooden bridge.

A red-headed man of slight build, with a long nose and a high forehead, Ithiel Town had stubby fingers that were constantly at work designing churches and public buildings, jobs which took him all over the country. While in North Carolina on a state capitol commission, he took time out to devise an entirely new type of bridge truss, which he called "Town's lattice mode." He took out a patent on it in 1820.

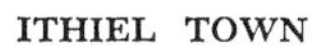
ITHIEL TOWN

The *Town lattice* truss was an uninterrupted series of crisscrossed diagonals in construction forming, as has been noted earlier, what were actually overlapping triangles. In such trusses any load on any one triangle affected distribution of stress in all other triangles. The web members were fastened at their points of intersection, so that independent action of any one triangle was impossible. Therein lay the great strength of the Town truss. It was a real invention, not resembling any design advanced for wooden spans in the thousands of years before its time that bridges had been built.

Ordinary pine or spruce plank was the usual material. Holes were bored in the planks to receive the wooden connecting pins used at every regular lattice intersection and at the places where the lattices were secured to the top and bottom chords. One competitor scoffingly termed Town's brain child a "garden trellis fence." The design was ideal for multi-span bridges, since each span did not have to be fitted individually to piers and abutments—as did arch bridges, for instance. Indeed, its proponents claimed that the Town truss "could be built by the mile and cut off by the yard." Because of advertising broadsides and word-of-mouth praise the Town truss became a favorite in the Northeast despite its light appearance. Agents for the sale of the patent rights flourished in every shire town, collecting a royalty from builders of $1 per foot of bridge to be built. Should the agent find an eager-beaver builder who had already put up a Town lattice bridge without payment of royalty, he would usually settle for $2 a foot.

Except for a few test jobs in the South and a little bridge at Whitneyville in Connecticut to introduce his invention into New England, Ithiel Town built very few covered bridges

Heavy planks, crisscrossed and pegged, made Town's lattice truss a distinctly American innovation. It became popular because its continuous web could be built by average carpenters.

himself. He was more the promoter, the gad-about salesman, who liked to pop up wherever a big new bridge was about to be built. Wining and dining the directors, he would deliver eloquent speeches in praise of his "mode," and induce local contractors to bid on the job and build with his plan. Copies of Town's pamphlets on building bridges with his patent method were spread far and wide. Gaining a steady and increasing revenue from the sale of royalties for his bridge, Mr. Town became well-to-do. He extended the use of his bridge design by doubling the quantities of plank and pins, and making the bridges larger and stronger to accommodate the locomotives of the early railroads. Bridge builders were still using the Town lattice truss long after Ithiel's patent rights ran out and there were no longer any Town agents nosing about the riverbanks trying to claim patent royalties.

In those early years the only engineering used in bridge building was by trial and error or rule of thumb. In 1830 Brevet-Colonel Stephen H. Long of the United States Army Topographical Engineers devised the first wooden truss in America into which a few mathematical calculations entered. It was a *panel* truss, which needed no arch. Viewed from the side the *Long* truss resembled a series of giant boxed X's.

Born in 1784 in Hopkinton, New Hampshire, Colonel Long returned to his birthplace on leave from the Army and established himself as a ranking bridge patentee. A tall man with soldierly bearing, he had a sharp and appraising glance. To some he seemed stuffy and austere, but there were merry eyes in his face—lined and weathered as it was from years of Western exploration—which gave hint of his cheerful and optimistic nature. Like Ithiel Town the colonel was more promoter than actual builder, continuing his experiments with wooden bridge trusses for thirty years. He wrote leaflets and a good-sized booklet which gave directions to bridge builders and offered suggestions as to the best procedures for erecting a patent Long truss. He made a model for exhibition that could be carried about in a small wooden box like a valise, and from Maine to Louisiana he appointed agents and subagents who were empowered to attend to the details of building bridges on the Long system.

This demonstration model helped to sell Long's truss as he and his rival scurried to win bridge contracts. Agents for panel or lattice designs ranged over the East and their efforts to publicize their spans compare well with today's big promotion campaigns.

His bridge enjoyed a popularity with builders for ten years, during which time it vied with Ithiel Town's lattice for favor among the growing railroad networks, toll-bridge companies and individual town highway planners. The rival bridge promoters exchanged polite notes via the newspapers, waxing almost poetic in description of their own designs. Colonel Long, busy with Army matters, finally left his share of the battle to his brother Moses. After 1840 the promotion of both Town and Long patent bridges was almost totally eclipsed by the advent of another bridge form. This was the *Howe* truss, invented by William Howe of Spencer, Massachusetts.

Detail of a Long truss.

STEPHEN H. LONG

Born in 1803, Howe, a millwright, had watched the Western Railroad being pushed through near his home and thought he could make a better bridge than the Long trusses that the road was building. Experimenting, he came up with a plan similar to that used by the railroad. The Howe design resembled Long's patent so much that the colonel claimed infringement for years. He never got any satisfaction from the patent office, for the "improvement" Howe claimed was a real step forward. In the Howe truss the upright wooden posts of Long's design were replaced by iron rods which could be adjusted with nuts and turnbuckles.

Thus William Howe was the first to cope successfully with the inherent weakness of wood in bridge construction: when it was used as a diagonal it could compress and, in doing so, increase its strength; but when it was used as an upright in a truss its joints pulled apart with the weight of loads passing over the bridge. With the advent of the Howe truss wood began to give way to iron for bridge construction.

Howe first tried his idea to hold up a church roof in Brookfield, Massachusetts. Then he got a small subcontract to build a railroad bridge over the Quaboag River in the village of Warren. This small span was so successful that Chief Engineer George Washington Whistler of the Western Railroad gave the experimenter the contract for the biggest bridge on the line, which was to cross the Connecticut River at Springfield. This one, too, proved to be a spectacular success.

WILLIAM HOWE

Howe came from a family of inventors, for his brother Tyler invented a spring bed, and his nephew Elias has lasting fame for his part in developing the sewing machine. A dreamy-eyed, handsome man with a fine tenor voice, William Howe's heart was in invention alone, and he delegated the building of his patent bridges to others.

They built with a will. Dozens of companies holding patent franchises built Howe truss bridges from Maine to California. They were in particular demand by the railroads, which needed thousands of bridges to carry tracks westward in the great expansion days of the new rail network. The Howe truss was equally popular as a highway bridge. Almost any good wood could be used—white oak, pine, fir, hemlock or cedar. Foundries in New England turned out the iron rods and nuts for Howe bridges as well as the inventor's patent angle blocks into which the ends of the timbers fitted. All the pre-cut timbers and iron parts for a Howe truss could be loaded on a couple of flat cars and shipped directly to a siding near the bridge site. Erection of the framework took only a day or two, with the covering to be put on at the builder's leisure while the bridge was in use.

These five New Englanders made the greatest contribution to covered bridge building in America. Palmer and Burr rediscovered the Palladian truss and arch designs and got things rolling by building important examples to follow. Town and Long put a substantial bridge design within the reach of the ordinary carpenter. Using a standard plan and unskilled labor, Howe brought the half-century of feverish building to a peak just before the Civil

The Gold Brook highway bridge at Stowe, Vt., is an excellent example of the Howe truss.

War, with hundreds of wooden spans being erected simultaneously. Many fine bridge builders came after these masters but their stories must wait to be told elsewhere in this book. Their names and bridges are noted in the Appendices, and their major contributions are described in Chapter V, the state-by-state account of the great covered structures of the Northeast.

The nation had profited greatly from the ingenuity of this quintet, but their influence was not confined to America. The success of their designs caused a resurgence of wooden bridge building in Europe—France, Russia, even the timber-poor British Isles used plans originated by these Northeast builders for their own railroad bridges. Incorporated into Europe's 19th-century highway bridges, the great trusses went, full-circle, back to Germany, Switzerland and Austria, where the very first wooden covered spans had originated centuries before. And present in countless steel truss bridges today is the legacy left by our early covered bridge designers. The modern, graceful steel arches and continuous-truss bridges over our largest rivers can be traced directly back to the plans in wood dreamed up more than a century ago by Palmer, Burr, Town, Long and Howe, and their followers.

Rare photograph of the first Howe truss bridge, built in 1838 for the Western Railroad (now Boston & Albany) at Warren, Mass., and replaced in 1873.

IV

Methods and Tools

A GREAT deal of ingenuity and sweat went into building a covered bridge. Unlike their modern counterparts, the men who made yesterday's bridges could not move mountains or streams to accommodate their needs. Our grandfathers had no such things as diesel shovels, bulldozers or dynamite, and the terrain—not men—dictated where the big spans should be. And when it came to actual construction their equipment consisted of a few crude tools, native skill and the pioneer virtue of not being afraid of a lick of work.

Only the largest bridges, at cities and major turnpikes, were left to be built and run as toll enterprises by private capital. When a bridge company was formed it usually solicited bids for the job, and men who made a business of erecting sizable bridges competed for the contract—each showing elaborate plans and models for the various trusses and expounding at length upon the superiority of the type of structure he proposed to build. The lowest bidder was awarded the contract and his crew of specialists went to work.

And specialists they were. Great innovators like Palmer and Burr, and lesser but still noteworthy figures like Nicholas Powers, supervised the erection of their masterpieces. In doing so they evolved streamlined systems for cutting and shaping giant timbers, having them ready when the abutments were finished, and arranging for roofs, siding and portals to be put on after the bridges were opened to traffic.

However, it was Long, Town and Howe who, through their agents and subsidiary

Documents like this lay in many a strongbox during the early 1800's when investors rushed to buy shares in toll bridge enterprises. A majority of such companies paid their stockholders well.

No. 21. Canajoharie, 26 January 1826

George G. Johnson is entitled to Two Share in the Capital Stock of the Canajoharie and Palatine Bridge company; transferable at the Office of the Treasurer of the said company: Subject to the further Payment of —— Dollars on —— Share, in such Sums and at such Times as shall be required.

(2. sh. of No. 5) . By Order of the Board of Directors,

John Frey President.

Henry I. Frey Treasurer.

builders, raised the correlation of labor and materials almost to the point of mass-producing bridges. Contractors using their patented designs would start by engaging gangs of men to fell and shape the necessary timbers. When the list of wooden parts was checked off skilled crews would be on hand to frame

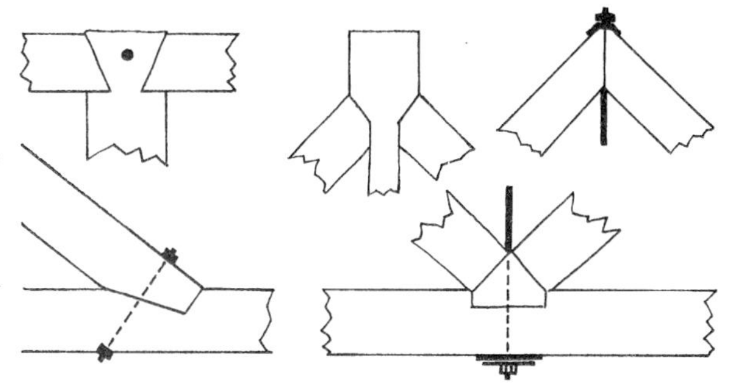

Typical joints used in wooden bridges.

the trusses on the ground or to erect them over the stream. If a Town lattice had to be snaked from one bank to another, a boss engineer would arrive to oversee this dangerous and tricky maneuver. If the truss was a Howe type, a load of iron rods, bolts, washers and angle blocks appeared at the scene in time for over-water construction to start without delay.

When the skeleton was securely in place the structural crews moved on, leaving carpenters to house the great beams and add sidewalks, buttresses, toll gates and whatever amenities in the form of paint and decoration were felt to be necessary. A heads-up builder might even take logs from his own woodlots, have them sawed to specifications at an adjacent mill and the lumber carried to the site of his newest job—thus anticipating the precut construction industry by nearly one hundred years!

For the smaller crossings, though, the towns themselves decided where their bridges would be and who would build them. Communities like Pittsford, Vermont, Swanzey in New Hampshire, Chester in Massachusetts or Salisbury, New York—all lying on both sides of a main stream—always seemed to have a petition before their town meetings for some new bridge to cope with the traffic of a single farm. Pittsford got its name from an early ford on the Pitts property. This was abandoned in favor of a covered bridge, built nearer to the growing village, at the Mead Farm. Hard by the Gorham and Hammond properties two more such bridges were thrown across Otter Creek, serving the northern and southern ends of town. When the railroad came along still another was built to give direct access to the depot. Three of Pittsford's Otter Creek spans are in use today, and are still designated as the Gorham, Hammond and Depot Bridges.

It was Avery Billings of nearby Rutland who went on record with the classic answer to a petition for one more bridge in his town:

"Mr. Moderator!" he bellowed in March Meeting. "Inasmuch as we have already built bridges over Otter Creek at Gookins Falls, Ripley's, Dorr's and Billings'—four bridges in two and a half miles—Mr. Moderator, I move we bridge the whole dang creek—lengthwise!"

For most New England towns building a permanent structure like a covered bridge was a major undertaking. A selectmen's committee usually took care of the details once the bridge had been voted. It set general specifications, obtained bids from local or outside builders, and in some cases even saw to it that the timber to be used was well-seasoned and reasonably free from knots. In New York State the town boards with their supervisors did the same thing.

It was a rare bridge that was laid out at anything but right angles to the stream. The accepted method of road planning called for a sharp turn into a bridge and one just as sharp at the other end where the traveler popped out. In the more leisurely days of horse-drawn vehicles this dangerous condition was tolerated, but with the advent of gas-buggies many an otherwise sturdy bridge was to be doomed by its right-angle location.

Ten men are dwarfed by a big railroad bridge as they pause in their work of roofing and siding. The picture may have been taken in New Hampshire in the '70's.

When the contract came to be awarded the local builder had a decided edge. His abilities were known to the selectmen or town board and the powers-that-be took into account the fact that his workmen would come from the immediate vicinity and that the wages they earned would therefore be spent at home.

The contractor had to marshal all his know-how for even these relatively modest jobs. He had to supply a working design for the bridge, see that the right sizes of lumber were prepared, select masons, carpenters, laborers and finishers. The number of men he employed might range from three to thirty, depending upon the length of the bridge. It was up to him to decide well in advance what means he would use to put up the span: some bridges had to be built from scratch, using falsework or scaffolding placed right in the stream bed; others could incorporate portions of an older bridge if one existed, and still others had to be framed section by section on land and the major parts hoisted out over the stream by block and tackle.

Meanwhile the abutments had to be built. There is more stonework in covered bridges than the casual observer might notice. Sometimes bridges were located on natural abutments of jutting rock and required a minimum of masonry—notable in this respect are the Brown Bridge in Shrewsbury, Vermont, and the Forge Bridge near Seager, New York. However, the great majority of Northeastern covered bridges have abutments of rubble masonry. This was made from unsquared chunks of stone in varying sizes, prepared for laying simply by knocking off weak corners and loose pieces. If the stones were fairly symmetrical they were placed in layers or *courses;* but for the most part no attempt was made to make the layers even until the top of the abutment was reached.

A few bridges had subfoundations of piling if the bottom of the river was unstable. A primitive horse-powered pile driver rammed long poles deep into the stream bed to provide a base for the masons to build upon. Occasionally, if a stream was deep and swift, it was necessary to build a small cofferdam around the abutment to protect it while the stoneworkers started the base; but more often the masons began their work well back from the normal water line.

They almost always laid stone "dry" for the abutments and piers of covered bridges—with the same technique used to lay walls—and the rock held tightly together. To bind the stone they often poured a thin liquid mixed from lime and cement into the mass of the abutment. This mixture, called *grout,* trickled between the stones, filling gaps, and hardened to make a solid piece of work. Mortar made of cement was a boon to the mason when it came into use. It was troweled on extensively to repair bridge abutments and piers, and the faces of the stones joined more solidly when they were "pointed up" with this binding material. Mortared abutments can be seen in covered bridges at West Arlington in Vermont, near Arena, New York, and at Robyville, Maine.

Rubble granite was used to a great extent in the states where it was quarried. Comstock's Bridge in Connecticut, the Moseley Bridge in Northfield, Vermont, and bridges in Winchester and Conway, New Hampshire, all have abutments of huge granite blocks laid in regular courses. Another quarry product, marble, found its way into the foundations of covered bridges: the Sanderson Bridge in Brandon, Vermont, and the Rexleigh Bridge near Salem, New York, have white-faced marble blocks, laid in regular courses called *range masonry.* But concrete or concrete facing has replaced much of the old stone masonry. The "modern" bridges at Hancock-Greenfield, New Hampshire, and Charlemont, Massachusetts, quite logically have new concrete abutments; in the Charlemont span they are faced with local fieldstone.

On any typical and well-organized job the workers in wood were busy while the stone substructure was being built. One of their first chores was to rig a catwalk across the stream, providing a precarious and roundabout footing from rock to rock. Sometimes carpenters put up scaffolding right in the river; and occasionally—if there was a special hurry to complete the bridge—they threw a staging together on the Winter's ice. This falsework was erected by making huge sawhorse-shaped frames called *bents,* then bracing several of these bents with diagonal timbers and laying a rude platform across their tops.

People who marvel at our early big bridges and at the ones in remote areas are paying an unexpressed tribute to the skill and integrity of the men who built them. Our respect deepens when we remember that hand tools were the only implements these men possessed for shaping, joining and finishing the great timbers.

First among these tools was the *broadaxe.* As the name implies, the broadaxe was a short-handled cutting tool with a broad, sharp

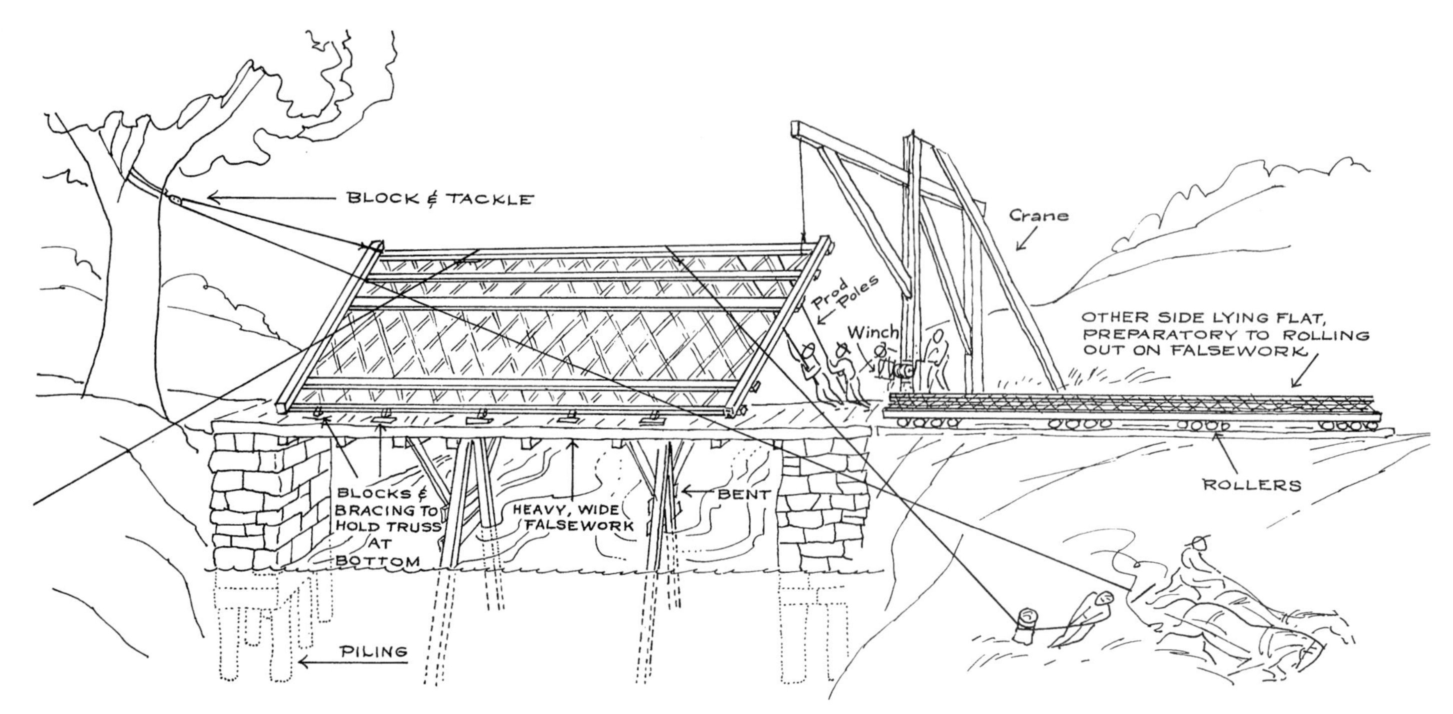

Town lattice trusses were raised across streams by two means: by men and horses—the only power available in early days of bridge building—or by wooden derricks, which came into use around the mid-century. Both ways are shown in this composite sketch. Old and young turned out to watch when the boss engineer organized the labor of inching the great webs over the falsework and joining them together.

blade, beveled on only one side. A regular axe cut down a tree, but the broadaxe hewed it from a round log into a square beam. The user notched the log at intervals, then hacked off chunks of wood between notches until the log had a flat, if rough, side. If an axeman was left-handed he simply reversed the head on the handle and started swinging. The beams that resulted were nearly as thick as the great pine and spruce trees from which they were made. Marks made in the hewing are clearly visible on many big timbers of the double-barreled covered bridge, more than a century old and now standing at the entrance to the Shelburne Museum at Shelburne, Vermont.

A different tool was employed for smaller logs and partly-finished timbers. This was the *adze,* an arched blade hung at right angles on a stout handle. If you have ever broken garden sod on a hot day with a mattock or grub hook, you have an idea of how an adze was used. A tall, thin man made the best adzeman. He would straddle the log, bringing the bright keen blade down between his legs and making the big chips fly as the wood took shape. A shorter one would stand on the timber itself

and hew right up to his toes. A good man could trim a flat surface almost as smooth as if he had used a plane, and finish by cutting a zigzag chunk from the end of the beam so that it could splice into another timber with a joint that was virtually waterproof. The adzeman excelled himself in wooden arch bridges, for it was his job to make precise joints so that the segments went together perfectly, and to cut the notches, or *daps,* for fitting the truss posts. If a timber was not naturally curved he could even hew a bit of a curve into it. The adzeman was proud of his skill and each had his own personal tool. Heaven help anyone else who so much as touched it!

The adze has been called the "Devil's own tool for danger." It's easy to see the hazard from a vicious blade, swinging rhythmically between the legs or within inches of the feet, where only a slight deflection could bring injury. Old-time adzemen bore the scars of their calling.

Once a cocky youngster came to a foreman on a bridge job, looking for work.

"What kin ye do?" queried the foreman.

"I'm an adzeman."

Up went the older man's eyebrows. He shifted his chew and looked the young fellow up and down.

"You be an adzeman, eh? Let's see yer legs!"

Caught, the boy sheepishly hauled his breeches out of his boots. His shins were as lily-white and unmarked as his face was red.

In the days when covered bridges were being built the mortise-and-tenon joint was the counterpart of the modern metal snap connection. One timber had a *mortise,* or broad slot, chiseled into its end. The connecting beam was fashioned with a flat tongue, or *tenon,* to fit snugly into the mortise. The early tenons were hewed out, but later a special tenon saw did the job. To strengthen the fastening a hole was bored through the joint and a round or square peg driven into it. In some cases a

Mortise and tenon made a slip-proof joint.

square peg in a round hole made the best and tightest connection!

This method of fastening wood brought special *chisels* into use. They were big and broad,

all-wood except for a sharp steel tip. They were pounded with a club-like hammer made of ironwood or oak called a *maul.* If you mauled a finger instead of your chisel you knew it in a hurry.

Holes were bored with an *auger*, an instrument like a great corkscrew with a wooden handle. For smaller holes a regular bit and brace were used, the brace being all of wood. The pegs were made in a variety of ways, usually by an older member of the building crew. Short lengths of wood were clamped in a shaving horse to be rounded, squared or pointed with a *drawknife*. Sometimes they were fashioned by driving the rough, unshaped piece of wood through a round hole bored in an iron plate. Still another method was to hew them carefully from sawed oblong "blanks" with a small broadaxe.

Walter Hard, Sr., the Vermont writer, tells of a workman who showed up one day when the Chiselville Bridge in Sunderland was being rebuilt. The foreman sized him up as being of no account and set him to work shaping pegs with a borrowed broadaxe, with a big boulder to use as a chopping block. The man worked steadily all day, his sharp blade not once touching the stone. Then, as the shadows began to lengthen and suppertime approached, the man lifted the axe high above his head and set the gorge to echoing as he shattered it on the rock.

"By Judast!" he exclaimed, "I was taught to put your axe in the choppin' block when you was done for the day!"

Saws were in use everywhere on a bridge job. The hand-, the buck- and the two-man crosscut saws still are much used today. Rarely seen now is the *pitsaw,* which got its name from being used in a pit. A log was placed over a pit and, with one man on a platform above it and another man standing below in the excavation, a long beam could be cut lengthwise—to the accompaniment of much sweating and heaving. The bottom sawyer tired in his shoulders and the top man tired in his back. If they changed places often during the day, come nightfall both were tired all over.

One more hand tool aided the bridge carpenter in his exacting labor. This was the *plane,* used to give hewn timbers more finish and smoothness. The size of the plane varied with the size of the man who handled it. One bridge builder's plane measured nearly three feet long and could send one wide shaving curling up and over the shoulder of the powerful young giant who used it.

Town lattice bridges presented special problems to the builder and his crew. Water-powered sawmills cut the wide spruce, pine and hemlock planks with their up-and-down gang saws. Coupled wagons hauled the lumber to the bridge site, and there it was eased to the ground with cant hook and crowbar.

If the adze was known as the "Devil's own tool" then the machine which bored holes for lattice trusses also warranted its nickname of "Satan's sure bet." This contraption was a

The boring machine in operation.

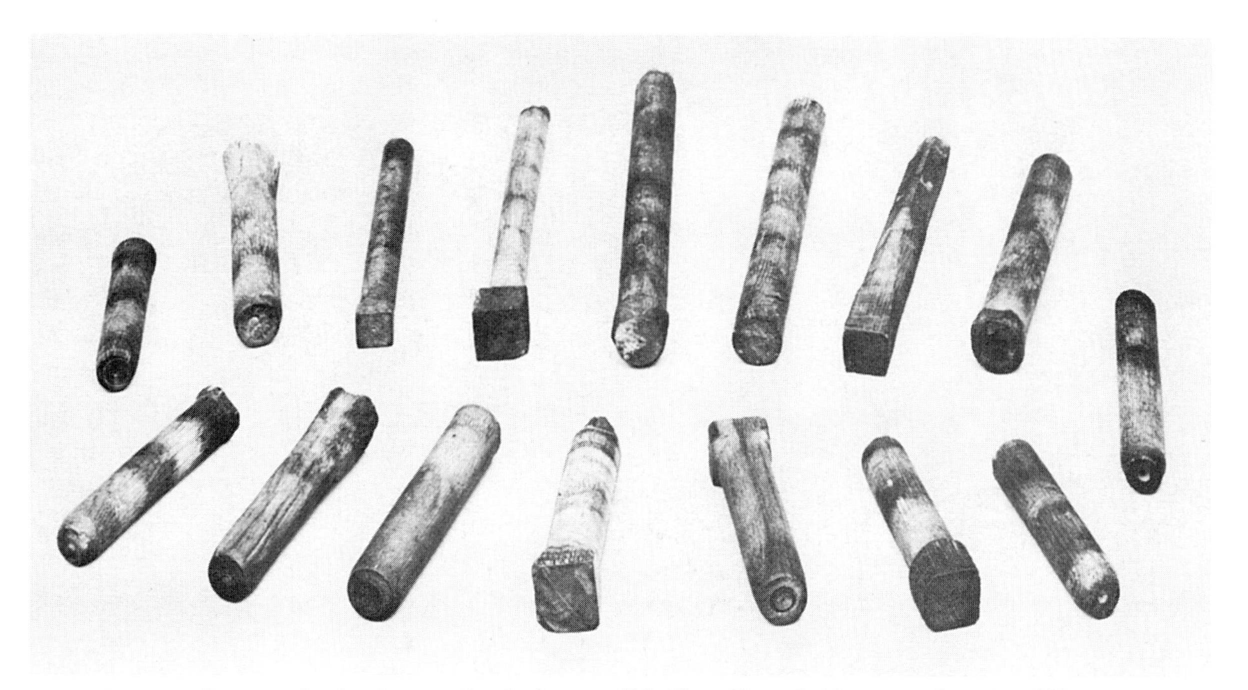

Some of the many kinds of treenails which are still holding Town bridges together after 100 years.

mechanical auger. The operator sat astride a plank, a little behind the boring machine and leaning over it. He located where the hole was to be and turned two crank handles which were geared to bore the auger into the wood. In this position he worked for hours, boring hole after hole from one end of the thick, heavy plank to the other. Often the machine would bind at a knot, screw blades would have to be replaced or sharpened, and the end of the plank always seemed miles away. Picture a man at work on a hot Summer's day, down in a valley where no air stirred, and faced with boring 2592 holes—the approximate number required to make both trusses of a 100-foot lattice highway bridge. He sweated and cranked away, denied even the zest of racing against a rival operator: boring these holes was a finicky business and spacing had to be exact.

If 2592 holes was an awesome prospect, think of making the 912 pins necessary to peg the planks together! (Or think of double this number of holes and pins—5184 and 1824, respectively—if it were a 100-foot railroad bridge whose extra sturdiness required double thick lattices in its trusses.) No wonder that builders of lattice covered bridges encouraged small woodworking shops to make oak pins, called *treenails* (and pronounced "trunnels"), which the contractors ordered by the thousand. Unlike the pegs shaped by hand on the spot to fasten massive timber joints, treenails were turned on a wood lathe. Then women and boys dipped them in linseed oil and tossed them into a big bin, where they waited to be taken to the bridge site by the wagonload. Look at the pins in most of the lattice bridges standing today and you will see the marks on the end of each round shaft where the lathe clamps held it more than a century ago. And because specifications and the work of lathemen differed so much from one bridge to another you can also find great variety in the pins—one to two inches in diameter and ten to twelve inches long, and short, fat, thin, round-headed, square-headed or headless—which are still holding the planks together.

Brawn and a sure eye took over once more when a lattice truss was pinned together. The planks for one whole side of the bridge were

An unusual picture from Dartmouth College archives showing a Town lattice railroad bridge being erected during the Winter. Note the height of the trusswork and the extra-heavy planking that was necessary to support the weight of a train.

laid out in the final crisscross pattern in a meadow adjacent to the abutment. Carefully, with a tap here and a nudge there, the holes were lined up to coincide, and the oiled treenails placed in the openings. Now up stepped a hefty individual to wield the *beetle*—pronounced "biddle"—a long-handled mallet with an ironbound oaken head. It gave off the satisfying clunk of wood against wood as the beetleman drove home the pegs and piece by piece the web became rigid.

At length all the treenails were in and the ticklish job began of jockeying the truss out over the river. All hands were called to help in this work, which called for plenty of rope and a thorough knowledge of knots. Men fell to, heaving with might and main. Shouts and orders filled the air, intermingled with cheers from the townsfolk who gathered to watch. Horses strained to hold lines snubbed tight around big stumps. Now the master bridge builder showed his worth as an organizing supervisor: in the final bridge raising everything had to move smoothly and according to plan.

Slowly the whole side of the bridge moved out on rollers over the falsework in the river, slowly it rose into place. There it rested while the other side of the bridge was urged into position; then the two trusses were joined by upper and lower bracing. Some builders gave the Town lattice truss a little upward *camber*, or hump, so that the floor would come exactly level when the weight of the plank settled on the oiled pins. It was a gala day when the bents and scaffolding were knocked away and the bridge stood alone.

Machinist's tools joined those of the carpenter when the Howe truss bridge came along with its iron rods, nuts and bolts. A good stout hammer pounded washers into place, and a

Looking downward on this panel one sees yesterday's bridge tools: froe for splitting shingles, chisels for gouging daps and mortises, two mauls, an auger and a broadaxe for shaping timbers.

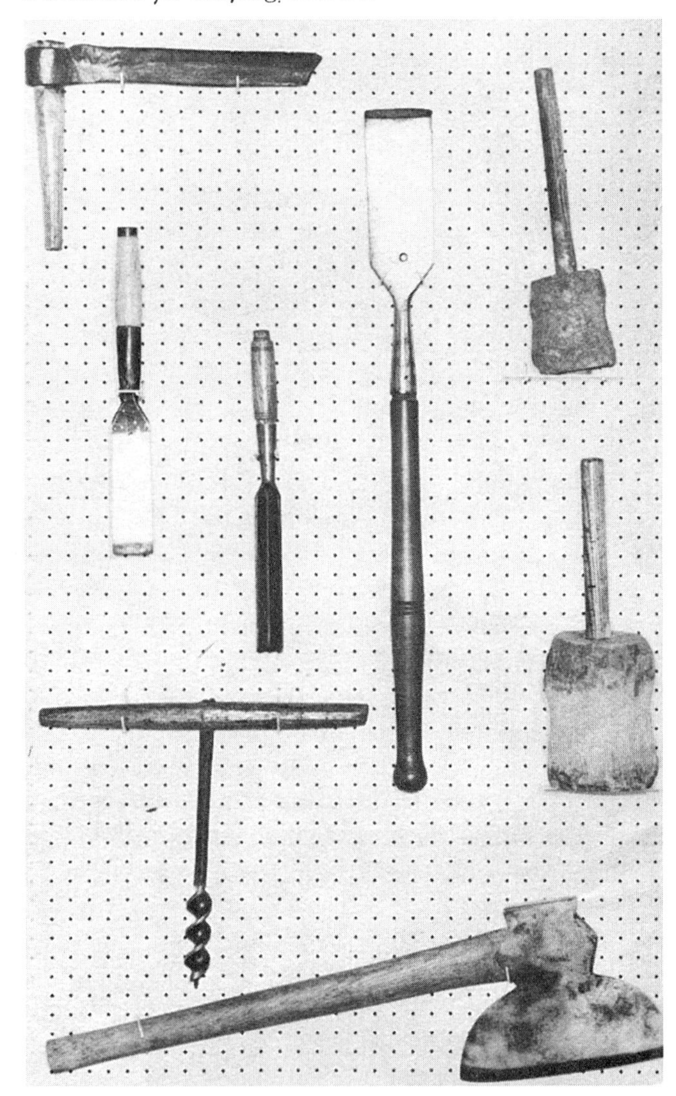

big open-ended wrench was enough to tighten a nut on the threads of a long iron rod. A first-rate builder of Howe trusses was careful to revisit his bridges within a year or so after construction to "tune-up" the spans. He and his crew examined the bridge carefully to find any sag or looseness. Then, where necessary, they tightened the turnbuckles and nuts of the ironwork and wedged the wooden fastenings together with mighty whacks of the maul or the beetle.

Arches of laminated plank were sometimes added to strengthen a bridge which had been strained by heavy loads. Iron rods were bolted through this laminated timber, and the floor beams relocated to hang from the lower ends of the rods. Arches of this type nearly obscure the originals in the Arthur Smith Bridge near Lyonsville, Massachusetts, and the Pulp Mill Bridge at Middlebury, Vermont. Such arches are also seen in some of the New Hampshire bridges, including those at Stark and Groveton.

Ordinary carpenters took over when the heavy construction work was done, and their first job was to lay joists and flooring over the roadway supports. As has been explained, the roadways of bridges were supported by the trusses, and were usually laid across between the lower chords. In an occasional arch bridge, though, the roadway was hung from the huge laminated bows and so did not rely at all upon the trusses which acted only as stiffeners for the arches. The wide floor plank was laid right-angled to traffic and was responsible for the clack-clack-clack so dear to the hearts of those who used covered bridges in their youth. The floorboards took a beating from heavy traffic and from the drag of sleighs and cutters as they scraped and squealed their way across the dry planking each winter. The answer was to "snow the bridges": the town hired people to draw snow into the bridges to slick the passage for the runners. Of course the dampness of melting snow played havoc with the planks, but flooring was the one part of a covered bridge which was most expendable and it was renewed every few years. Today most bridge floors have been replanked lengthwise, or at least have parallel runners that offer smooth going for rubber tires.

Only the big pretentious covered bridges had divided lanes. In most of the rest there was just about room for two rigs to pass. Once in a while a sign admonished drivers to KEEP TO THE RIGHT AS THE LAW DIRECTS, but for the most part it was a case of first come, first through. Yet some covered bridge floors were simply not wide enough to allow two teams to pass each other. And then some choice profanity issued from under the eaves! It was no mean feat to back a pair of high-spirited horses and a four-wheeled wagon out of the narrow passage, and many a time traffic stopped for half an hour while adamant drivers, jammed head-on in the middle of a bridge, sat and glared and refused to budge.

The bridge opened for traffic as soon as the roadway was laid, and it was then that the carpenters set about adding the roofs and the siding. They were trained in raising roofs for their houses and barns, so the lighter carpentry went ahead speedily. Sometimes they felt that additional bracing might be necessary to give sturdier resistance in a windy valley, and they inserted naturally-curved wooden pieces called *ship's knees* to strengthen the right angle between the sides and tops of the bridges.

Each Christmas Santa Claus and his reindeer come to rest on the ridgepole of the "retired" bridge at Groveton, N.H., now almost 130 years old.

Regular gable roofs were fairly standard for covered bridges, but there were exceptions such as the flat roofs of bridges near Dover, New Hampshire, and Rawsonville, Vermont, and the hipped roofs of large spans in New Hampshire and Maine. Another variation was to roof only the trusses themselves. A wooden bridge with low trusses which had been roofed and enclosed appeared at first glance to be merely an open stringer affair with remarkably heavy railings. Rarely did a highway bridge have a cupola on the roof for illumination but a number of the best railroad bridges had smoke vents in their coverings.

Good shingling was a necessity if the bridge was to last. Shingles were split from a block of dry pine or cedar with a tool called a *froe*. The froe was rather like a wide, shallow wedge with a handle set at a right angle on the end of the blade. The user held the handle upright and struck the top of the blade with a maul. Off would split a nearly-finished shingle called a *shake*, much like a thin pie-shaped piece cut from a wheel of cheese. Hundreds of shingles went on the roof of a covered bridge and they were usually purchased by the bundle from independent makers.

How clapboards were split from a log for siding.

The very early covered bridges, like those in the Town of Bennington, Vermont, were sided horizontally with narrow clapboards rived from big dry logs. Later, a few bridges were sided with shingles, but the usual material was thin planking nailed on vertically with the seams sometimes protected by thin battens. The builders used siding as an additional protection to the trusses, so they didn't worry very much about whether or not a traveler had a clear view of oncoming traffic. A majority of highway bridges—and virtually all railroad spans—had solid siding clear up to the eaves. But here and there a builder had a feeling for design and laid on his siding to indicate the sweep of a great arch; or perhaps he sided his bridge only up to eye-level, partly to show off the trusswork and partly to give drivers a chance to avoid meeting another rig in mid-span. In modern times, when bridges needed to be re-sided, a portion of the truss near each entrance might be left without siding to give automobiles a more-or-less clear view of the access roads. It is doubtful if builders left open spaces in the siding in order to reduce the impact of high winds, but it was fairly common practice for a daredevil crew to rip off the plank sides of a bridge to give flood waters a clear sweep through the trusses and so keep the structure from being carried from its abutments.

A few spans in Vermont and New Hampshire, and in Delaware and Sullivan Counties in New York, have an unusual wrinkle in covered bridge building. This is the *buttress*, a small covered triangular brace protruding from the outsides of the trusses. In a head-on view of the bridges these buttresses have been compared with the droopy ears of a sad beagle hound.

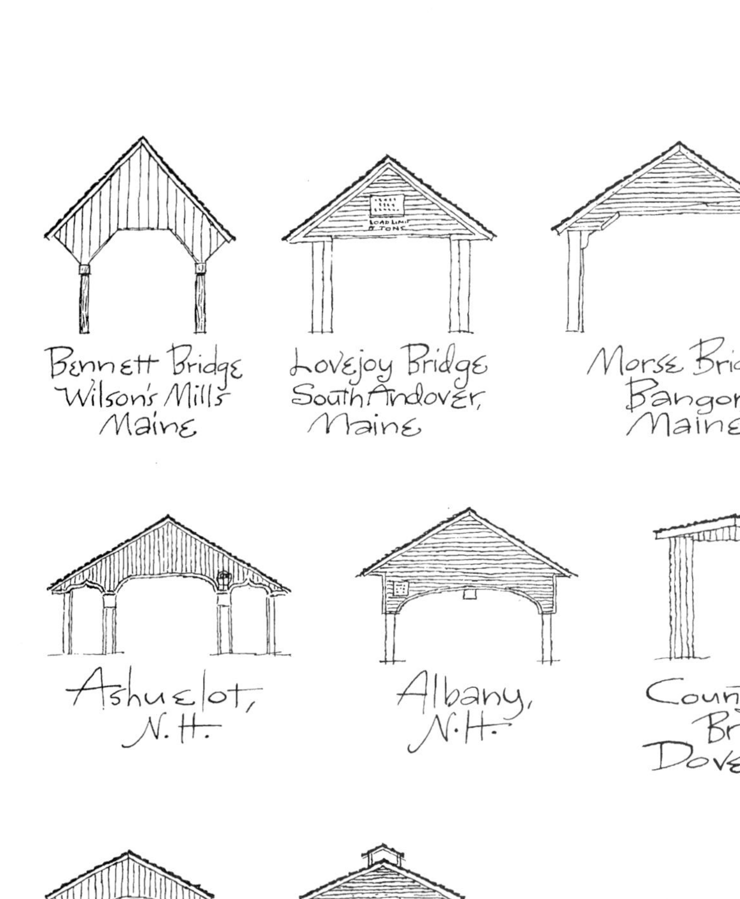

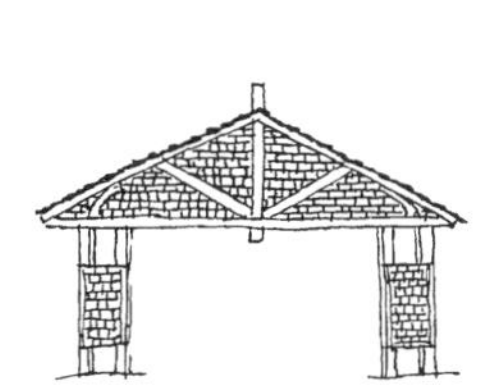

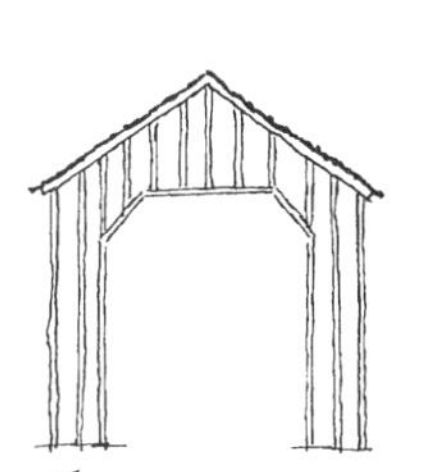

SOME REPRESENTATIVE PORTALS, 1957

The portals, or entrances, were the only places for the builder of a covered bridge to show his originality where it could be admired by all comers. Practical and unostentatious, New England contractors seldom took advantage of this opportunity, and in general bridge portals in Yankeedom are plain to the point of severity. Yorker builders were less inhibited, however, and some of their bridges sported portals with painted eagles, dates, mottoes, and the names of those who projected and built the bridges. This freedom survives in the name stencilled in an arc across the portals of the Hoosic River bridge at Buskirk.

Many towns, like Winchester, New Hampshire, and Montgomery, Vermont, kept the interiors of their village bridges scrupulously whitewashed, both to preserve the wood and to retard fire. This gave the bridges a rather ghostly atmosphere when one traversed them late at night.

Paint? Most covered bridges never saw a drop until modern times. The builders figured that well-seasoned siding didn't need it, and new siding was real cheap, anyway. There were exceptions, of course. Bridges in Brattleboro, Lyndon and Bennington, Vermont, in Woodsville and Swanzey, New Hampshire, and in Eagleville and Newfield, New York, have always been painted barn red, back as far as anyone can remember. The new span at Sheffield, Massachusetts, carries on the "old red bridge" tradition. White paint shines on bridges at Ashuelot, New Hampshire, and Sergeantsville, New Jersey. Some, like the big Windsor-Cornish Bridge between Vermont and New Hampshire, have been repainted a neutral gray. The former Holland Bridge in Townshend, Vermont, was once completely doused in bright green! Right after the Civil War the Salisbury Station Bridge over Otter Creek in Vermont was made resplendent by bright yellow ends trimmed with red, and traces of this décor are still visible. In Northfield in the same state several bridges have candy-striped portals—red, with the battens at each joint of the boards painted white.

The most striking example of painting in recent years was found in the single-span bridge over the Ware River at Gilbertville, Massachusetts. The bridge is jointly owned by the Towns of Hardwick and Ware. In 1954 Ware decided to spruce up its half of the bridge and had all new siding and roofing put on, right out to the middle and no farther! The renovated portion was painted a good barn red. There the bridge stood, half refurbished, half decrepit and neglected. Hardwick eventually got around to repairing and repainting its half of Gilbertville Bridge, but with weathering, the dividing line will always be discernible.

Gilbertville Bridge after Ware, Mass., renovated its half.

V

State by State

NEARLY two hundred covered bridges still stand in the Northeast, a fact that will come as a surprise to many. But it shouldn't: this region was the real birthplace and proving ground for our American covered bridges, and it is here that they have been in use for the longest time. Nearly all of the foremost innovators and builders were either born, lived or worked in the Northeast: and, although a great number of their historic bridges have disappeared, many examples of their handiwork remain.

MAINE: Wood to Spare

MAINE, THE LARGEST New England State, nudges its bulk far up into the Dominion of Canada. A rugged people live in the valleys, on the seacoast and along the level potato country in the northern tip, but a good share of Maine is a vast rocky wooded tangle of lakes, rivers and mountains.

State of Mainers built a good many covered bridges across the rock-strewn rivers which come tumbling down from the mountains to the sea. Maine was primarily a lumbering state, and every good-sized falls had its sawmill, with a sound of whining blade and a pungent smell of sawdust in the air. It was also a country of shipwrights and farmer-mechanics who had the know-how to put a covered bridge together. In fact, with all the sawed wood available, it is surprising that more covered bridges were not built in Maine. The hundred and twenty or so of which there are records stretched in a long crescent from the New Hampshire border to the Canadian line. Many sections of the state were hastily developed for timber; they were thoroughly exploited and then left for more gradual settlement. Only the most important arterial highway crossings at sizable towns were thought worthy of a stout structure like a covered bridge.

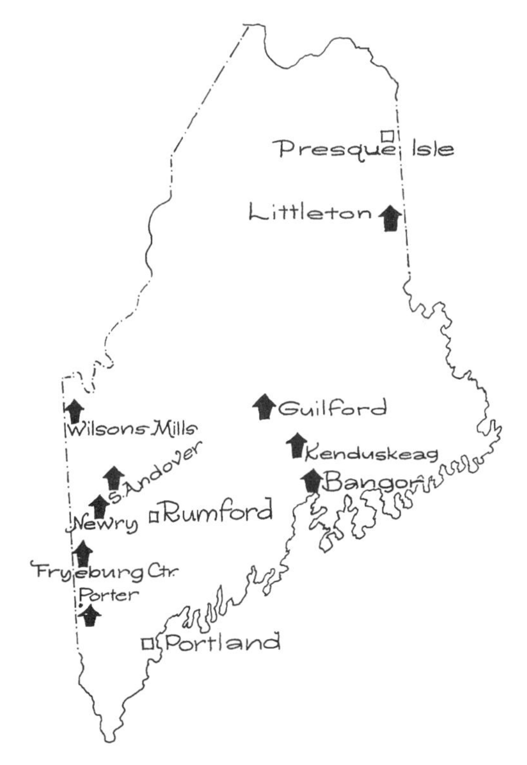

This view of two covered bridges over the Kennebec at Waterville—one for a highway and the other for trains—shows how towns along Maine's principal rivers used to look years ago.

Maine's first substantial bridge was probably the big one at Augusta, spanning the Kennebec River. The original bridge on this site, an open structure built by Timothy Palmer, was put up by a private company in 1797 when Maine was still a district of Massachusetts. The covered bridge which replaced it, thought to be the first of its kind in the state, was erected in 1819.

It has been said that Lieutenant Jefferson Davis, while on duty with the United States Army Topographical Engineers, supervised the erection of a long covered bridge at Mattawamkeag, Maine. The young Southerner may have been in Maine in connection with the building of the Military Road from Bangor north, but careful research shows that the Mattawamkeag Bridge of 1831–1927 was built under the direction of another army engineer, Captain Charles Thomas.

Checking records against a map shows that the great majority of Maine's covered bridges were in the Saco, Androscoggin, Kennebec and Penobscot Valleys. There were big ones equipped with sidewalks for pedestrians at well-known places like Old Town, Skowhegan, Waterville, Gardiner and Biddeford. At Dover-Foxcroft the bridge's footwalk entrances were embellished with gothic arches at the portals, making the span an attractive haven for Sunday strollers. Bangor Bridge, stretching 792 feet across the Penobscot to Brewer, was the longest covered structure ever built in the state. It had portals with flared tops, reflecting a craze for Egyptian architecture which swept the country at the time it was being built in the 1840's.

Early pile and wooden crib abutments proved poor economy in the yearly Spring freshets, so beneath most Maine bridges were well-built granite block foundations. Also found in several Maine towns (notably in Oxford) were wooden spans whose roofs and sides were completely shingled, a siding unusual in the East although common in Oregon and California. The Robyville Bridge, still standing northwest of Bangor, is the sole surviving shingled bridge in the state.

Over the years Maine residents made good use of their bridges. Village spans with light traffic were often used as shady places to hitch a team or two while their owners bargained in the nearby general store. They were a natural place for young and old to watch the annual log drives. The lumber companies often stationed a man out on a projecting bridge pier to keep the logs moving. Peavey in hand and jaws full of tobacco, he would stand like a ship's lookout, eyeing the flow of logs and alert for a pile-up.

The unusual old double-barreled bridge which stood over the Stillwater River at Stillwater served yet another purpose. Not long after the bridge was completed in 1836, the village residents had to decide whether they wanted to become part of Old Town or of Orono. The new bridge, dry and convenient, served not only as an assembly place for the meeting but as an effective polling mechanism. All the male voters who wished to unite with Old Town were told to gather for a head count in one lane of the bridge, and those favoring Orono to stand in the other. Old Town won by a majority of about twenty-five votes. One man on the Orono side, who saw how the vote was going and didn't want to be among the losers, is said to have crawled through the lattice truss partition in the middle of the bridge in time to be counted with the winners. He very nearly lost his trousers and his dignity in the process. At the time it was replaced in 1951 the Stillwater Bridge was the last two-lane Town lattice truss covered bridge in the United States.

Log jams, fire, flood, ice and progress all took their toll of Maine bridges. Floods always were the worst enemy, and town histories are full of terse reports, such as: "Bridge built in 1829, Carried away 1832, Rebuilt 1835, Damaged 1836, Rebuilt, Out in flood 1869, Rebuilt 1870"—and so on. Always present was the fear that, even if the local bridge was strong or high enough to withstand the freshets,. some other bridge from upstream would break loose and come floating down to wreck it.

In fact, in high water in October, 1869, the Ticonic Toll Bridge at Waterville, Maine, floated off its abutments and started down the Kennebec. The city officials at Augusta, 19 miles downstream, got the news by telegraph. Fearful of what the derelict would do to their own bridge, they hastily organized a crew of husky men armed with ropes and grapples,

P. T. Barnum's Jumbo tested this sturdy international bridge before leading the circus across from St. Stephen to play Calais. The toll and custom house stands at the entrance to the span from the U.S. side.

One of the few covered railroad bridges with a swinging draw spanned the Penobscot at Bangor.

chartered a train and set off upriver. Drivers churning, the little engine and its anxious load sped alongside the flood-swollen river, the men expecting at every bend to see the broken Ticonic Bridge bearing down upon them in the current. Luckily, they found the bridge snagged at Vassalboro, and got two volunteers out to her with ropes. Backing the train slowly, they maneuvered the battered, half-submerged hulk into a quiet reach of the angry Kennebec to sit out the crest of the flood. The Augusta bridge was saved.

Rivers separated many Maine towns, and road commissioners argued right down to inches over the cost for building and maintaining the covered bridges that linked them. At West Buxton many years ago, Hollis and Buxton shared such a bridge across the Saco River. In time it was replaced by an iron bridge. The latter went out in a freshet, as did the temporary bridge which succeeded it the following year. When it came time to rebuild, Buxton erected the half for which it was responsible in iron. Hollis, however, would have none of this: it favored the old-fashioned, tried-and-true wooden covered bridge. And for nearly forty years thereafter the West Buxton Bridge spanned the Saco, built half of wood and half of iron. Engineers kept an eye on the structure, curious as to the relative strength and longevity of the two halves. When the great ice flood of March, 1936, came rumbling over the dam at West Buxton, it removed all such speculations. The ice took out the whole shebang, smashing wood and iron trusses indiscriminately.

Downstream on the Saco another covered bridge, the Union Falls span between Buxton and Dayton, had an even more dramatic finish. The Saco valley at this site had been chosen for a new power dam and the bridge had to go. At that time Pine Tree Pictures, Inc., was making six-reelers in Portland, and the destruction of the bridge would fit in handily with a silent movie they were making from one of James Oliver Curwood's stories. Eyes gleaming, the movie director made a suggestion to the power company:

"Let's blow it up!"

November 3, 1921, was a regular Roman holiday. Former Governor Carl E. Milliken, who was associated with the film company, was there, as were town officials, cameramen with their caps turned backwards, and an assortment of actors who did stunts on the bridge and showed off their dramatic talents before the big crowd that had come out from Portland. Sixteen sticks of dynamite and ten pounds of powder were placed in the middle of the seventy-year-old bridge. Shortly after noon the blast was touched off. The covered span rose high in the air, broke in two, and slumped in final glory into the river while

The only known covered bridge built expressly for trolley cars crossed the Souadabscook at Hampden. Portal design allowed room for power lines.

Hemlock Bridge over the old channel of the Saco near Fryeburg is worth the dusty ride to see it.

the cameras turned. Then it was soaked with gasoline and set afire. In the excitement one man exuberantly leaped into the Saco and had to be dragged out. By three o'clock the moviemen had had their field day and old Union Falls Bridge was only a memory.

Another showman, a four-footed one, used the old international covered bridge that crossed the St. Croix River from St. Stephen, New Brunswick, to Calais, Maine. Jumbo, the famous elephant and biggest drawing card of P. T. Barnum's "Greatest Show on Earth," came to play St. Stephen and its American neighbor. The story goes that Jumbo entered the gloomy cavern of the bridge, testing it gingerly with his huge feet. In some kind of elephant snort he informed his keeper that the bridge was safe. Then he led the whole parade of other ponderous pachyderms (Barnum's words), wagons with caged lions and tigers, and the brassy steam calliope through the bridge for a triumphant entry into Calais. The customs men on the Maine side did not attempt to make the elephants open their trunks!

Jumbo ordinarily traveled by rail, and the early covered railroad bridges of Maine were built extra strong to sustain such tonnage as Jumbo's, as well as the weight of locomotives and trains loaded with pulpwood bound for the mills. There were long covered railroad bridges across the Kennebec at Waterville and at Madison; and the one at Bangor had a draw so that high-masted schooners could dock farther up the Penobscot. In Phillips, the Sandy River & Rangeley Lakes Railroad maintained a covered bridge over the Sandy River for their little trains. This was the next-to-the-last bridge of its kind in the United States on a narrow-gauge railroad (the last was in Tennessee). Tops in rarity was a little one-span structure over the Souadabscook Stream

Sunday River Bridge near North Bethel, a favorite with artists, has been saved under Maine's preservation program.

in Hampden, Maine. This was the only known covered bridge in America to be built solely for the use of trolley cars.

From a high of some hundred and twenty covered bridges, Maine is down to just nine today, but they are a fair cross section of the types that used to dot the countryside. Farthest north of any covered bridge in the United States is the Watson Settlement Bridge over Meduxnekeag River on the road from Littleton to Woodstock, New Brunswick. The Morse Bridge near Valley Avenue in Bangor is the longest (212 feet) covered bridge in Maine, and the only one remaining today

Robyville Bridge is the sole survivor of the many covered spans in Maine which had shingled portals and siding.

within the limits of a New England city. A few miles up the Kenduskeag is the shingled Robyville Bridge, and farther north, between the towns of Sangerville and Guilford, the long bulk of Lowes Bridge over the Piscataquis dates from 1857.

Built that same year is Hemlock Bridge over the old, by-passed channel of the Saco River on a little used dirt road in Fryeburg. On the road to the old lumber settlement of Ketchum, in the town of Newry, stands the Sunday River Bridge. Artists portraying this bridge have daubed more paint on canvas than has ever been slapped on its weather-beaten sides. Not far away is the Lovejoy Bridge at South Andover, recently renovated and proudly showing signs of local attention and affection.

Bennett Bridge spans the Magalloway River south of Wilson's Mills in the Aziscoos Lake region near the New Hampshire line. Tucked away in the country just a dozen miles from Portland was Babb's Bridge. Back in 1864 it cost the adjoining towns of Gorham and Windham only $318 to build. Vandals burned it in 1973 and the price of its replacement will be over two hundred times that amount.

Porter Bridge over the Ossipee River completes the roster of present-day Maine covered bridges. This is another crossing under a two-town setup, being shared by Porter and Parsonfield; and as usual the old bone of contention was just how much of the structure each town was responsible for. When the present bridge was being planned a century ago, a squabble started. It was argued that Parsonfield would get more use from the bridge than Porter would, and that therefore the former should pay more for building and upkeep. The debate waxed hot and heavy and was getting nowhere until Ivory Fenderson, selectman from the south side, suggested that the disputing parties meet at the site in question. The selectmen stalked out on the old bridge, boots clumping on the sagging floorboards, and met approximately in the middle. Ivory stopped, pulled out his jack-knife and pitched it with a flourish into the plank at his feet. Then he announced solemnly:

"The Town of Parsonfield shall build so far, and no further!"

Ivory Fenderson's ghost rests easy, for his invisible knife mark still divides the crossing, and today Parsonfield and Porter share equally in the upkeep of their covered bridge across the Ossipee.

Bridge over the Ossipee River between Parsonfield and Porter.

NEW HAMPSHIRE: *Granite Base for Wooden Spans*

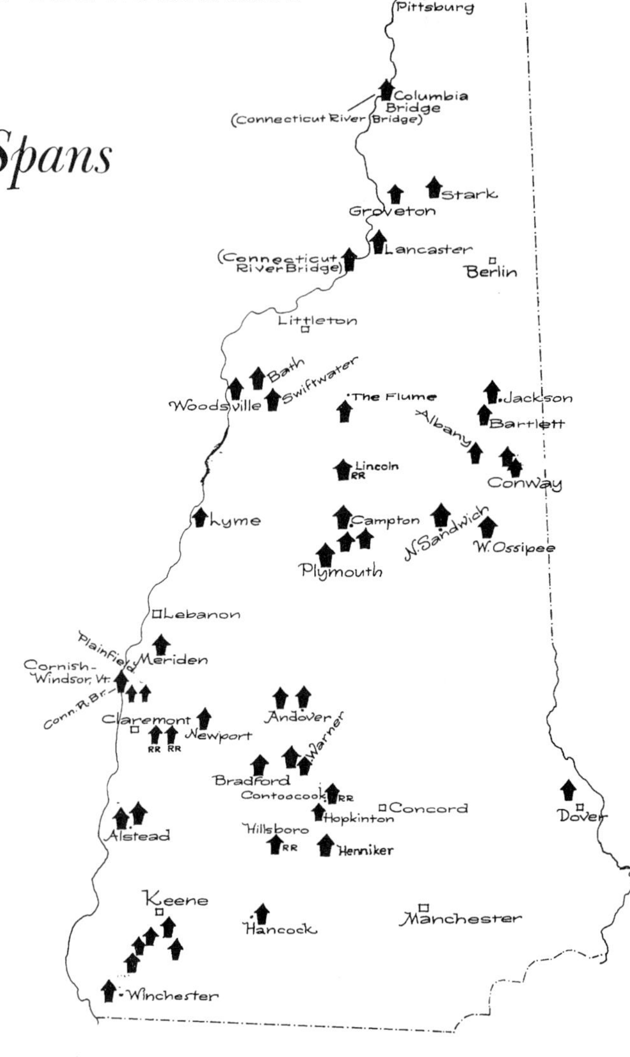

THE GRANITE STATE managed very well for a number of years without covered bridges. The big ones at much-used crossings such as Newington-Durham, Manchester, Concord and Portsmouth were all open truss spans to begin with, and the replacements often made necessary by the annual Spring freshets were of similar design. Colonel Enoch Hale is justly celebrated as the builder of the bridge at Walpole, New Hampshire (to Bellows Falls, Vermont), the first bridge of any kind over the Connecticut River. Erected in 1785, Hale's bridge used a natural outcropping of rock as a center pier and apparently had an open deck roadway with timber underbracing. Details concerning its trusswork (if any) have long been confused with that of a 10-span stringer bridge on wooden piling which replaced it in 1798.

Most bridge architects of that early day did not attempt any world-shaking arches or involved trusswork, but nevertheless bold schemes were being advanced. Rufus Graves, proposing to cross the Connecticut from his native Hanover, New Hampshire, to Vermont on the far side, made a model of the bridge based on Timothy Palmer's Great Arch of the Piscataqua down Portsmouth way. A promoter of no mean ability, he solicited subscriptions for stock in a company to build the full-sized span, and exhibited the model around the countryside to show what he was up to. In 1796, when he had raised enough money, Graves proceeded to erect the bridge, a huge uncovered wooden arch across the Connecticut, 236 feet in a single span. Unfortunately, though, he lacked much of Palmer's engineering talent, and eight years later the big bridge fell with a crash that echoed for miles up and down the river.

Covered bridges are not recorded in New Hampshire until the 1820's. Among the first to be erected is one still standing at Woodsville, built in 1829 on Ithiel Town's lattice plan. It was a forerunner of more than two hundred which once dotted the landscape of the Granite State and whose profusion prompted the local sarcasm used when somebody came into a warm room and left the door open:

"Were you brought up in a covered bridge?"

Colonel Stephen H. Long and his bridge truss were a big influence in his home state. Bridges on Long's design were thrown across the Contoocook in Bennington, Hancock, Hillsboro and Henniker. Top-flight agent for the patent was the colonel's brother, Dr. Moses Long of Warner, New Hampshire. Dr. Long, also the town postmaster, used his free franking privilege to blanket the Northeast

Tucker Bridge over the Connecticut between North Walpole and Bellows Falls, Vt., stood for 90 years.

with advertising broadsides extolling the virtues of the Long truss plan. Across the Connecticut River between North Haverhill and Newbury, Vermont, stretched New Bridge, the prototype used by the colonel and doctor to point out the superior advantages of the Long truss to journeymen carpenter-builders. The Childs brothers of Henniker, New Hampshire, erected New Bridge in 1834. Nearly eighty years later it was one of the oldest covered spans across the Connecticut, but when it succumbed to the flood of 1913 it was still called New Bridge locally.

The three Childs boys, related by marriage to the Long family, were the busiest and best-known of all the Long subagents. The oldest brother, Horace, was master carpenter. Enoch, a Yale graduate, made the designs and handled business details, while Warren Childs, a mason, attended to the all-important stonework down below. Here was a happy combination of talents working to perfection in dozens of bridges which the brothers built on the Long plan. In later years Horace Childs formed his own company to construct railroad bridges and secured contracts from Maine to Connecticut. He patented a truss of his own design; but, oddly enough, the only bridges on record built according to his plan were (and are) in far-off Preble County, Ohio! Childs bridges of the Long type in New Hampshire spanned the Connecticut at Lebanon, the Merrimack at Hooksett and Manchester, and the Contoocook at Henniker and West Henniker.

Enoch Childs did not confine himself to Long bridges: he modeled an experimental structure on the patent of Daniel C. McCallum of New York City. This was an unusual affair with curved upper chords and a correspondingly curved roof which won it the nickname of Rainbow Bridge. It arched over the Merrimack River between Boscawen and Canterbury, New Hampshire.

Farther down the Merrimack, the best-known of Manchester's covered spans was the Old Amoskeag, a toll bridge of arch-truss construction built in 1854 over the falls at the upper end of the city. Dow's, a local store, advertised boots and shoes in a display adver-

The charred skeleton of Boscawen Bridge over the Contoocook near Penacook still carried foot traffic for several months after being condemned.

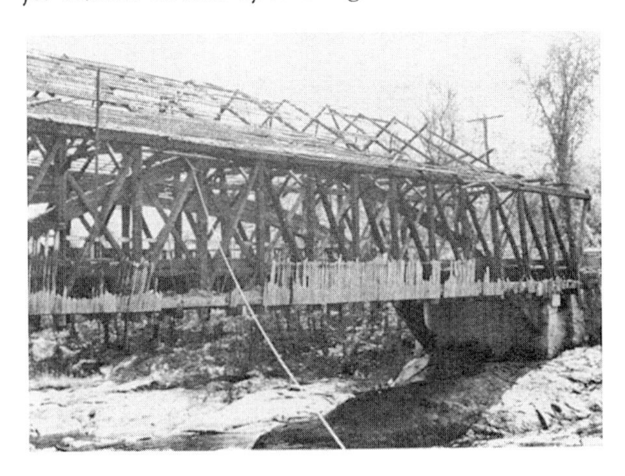

A century ago the Town of Conway was finagled into buying Joel's Bridge over the Saco to Redstone. In 1934 the staunch Paddleford trusses were strengthened by new splicing and ship's-knee braces hewn from crooked tamarack timbers. Burned by arsonists, July 5, 1975.

tisement painted the length of the upstream side of the bridge where it would surely be noticed by shoppers coming to town on the steam cars. To avoid a bad railroad crossing at the eastern entrance, Old Amoskeag was built up in step-like sections with an inclined ramp over the tracks.

A builder on the Long truss plan who successfully evolved a design of his own was Peter Paddleford of Littleton, New Hampshire. Paddleford had put up good Long bridges on the upper Connecticut at Northumberland and Monroe, as well as a fine one with laminated arches at Plymouth, New Hampshire. Then he modified the plans and stiffened the trusses by superimposing panels. The result was the unpatented but widely used "Paddleford bridge." The first known Paddleford bridge is thought to be the one which stood between Redstone and Center Conway, New Hampshire. Others were built at Conway, Concord and Penacook.

Lyman Bridge at West Lebanon.

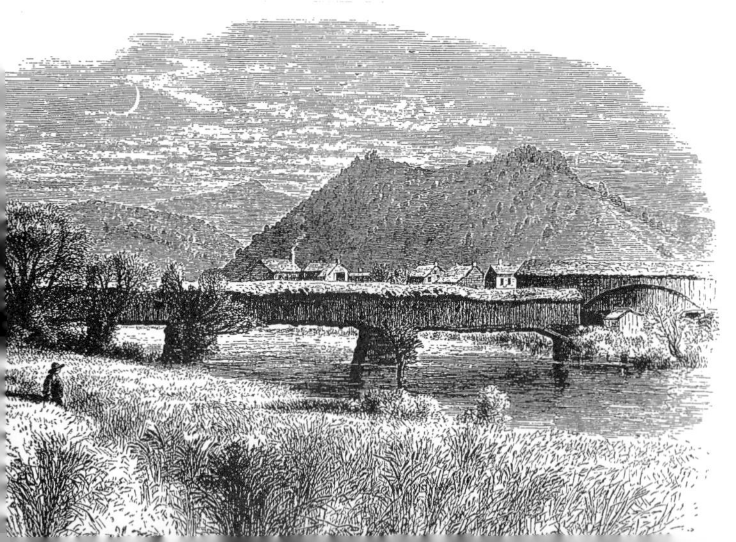

Paddleford built a bridge over the Saco River at Conway with the assistance of Jacob Berry. Berry became a convert to the Paddleford plan and is responsible for others of this type in Conway, Ossipee and Sandwich. The tradition went on to the Broughton family—Charles and his son Frank—who built the present day Conway bridge over the Saco River in 1890. Peter Paddleford's own son, Henry, a dour individual with a bald head and black beard, put up covered bridges in the Littleton area, using his father's plan and those of others.

By federal decree the Connecticut River is owned by New Hampshire right up to the low-water mark on the Vermont shore. So Vermont had to pay for only the small portion —usually from three to fifteen feet—of any bridges extending beyond the west bank. Thus New Hampshire has always had the lion's share of jurisdiction over the Connecticut River spans. Although a Vermont town might have been eager for a bridge, it was New Hampshire that had to pay for and maintain it after the company operating it dissolved and the bridge became free.

At one time there were thirty-five covered highway and railroad bridges across the Connecticut between its source in the lakes of Pittsburg, New Hampshire, and the Massachusetts line at Hinsdale. These ranged from a flimsy farm bridge below the first Connecticut Lake to the twin structures across the wide river at Brattleboro. Among the best-known were the Woodsville-Wells River combination railroad and highway span, the two quite separate Lyman bridges at Monroe and Lebanon, and the Tucker Bridge at Bellows Falls. All of these were once toll bridges, originally built with a view to lining the pockets of the investors.

An exception was the Ledyard Bridge between Hanover, New Hampshire, and the railroad station at Lewiston, Vermont. Old Dartmouth men have fond memories of the Ledyard span, for generations of undergraduates used it going to and from the college. And, of course, the bridge was named for the Dartmouth hero and world traveler, John Ledyard, who hollowed out a canoe from a tree near the site and began his celebrated journey with a voyage down the Connecticut. The fourth Ledyard Bridge on the same location was built in 1859; it was the first free bridge across the Connecticut River, being paid for by the adjoining towns and by Dartmouth College. Professor Sanborn delivered a dedicatory address at the opening ceremonies, at which he declaimed: "Let no vandal hand be raised to deface this noble structure, or injure one fiber of its timbers. Palsied be the hand that shall aid in its demolition. . . ." It is probable that none of the wrecking crew who tore down old Ledyard in 1934 ever read Professor Sanborn's address with its fervent threats; the remains of the old Dartmouth landmark might otherwise still be there.

Ledyard Bridge at Hanover was Dartmouth's pride and the first free span across the Connecticut.

Oddities among early New Hampshire covered bridges included the so-called Repub-

Casual vandalism derails the proud Atlantic *and breaks the back of a railroad bridge near Lebanon in 1883.*

lican Bridge spanning the Pemigewasset River at Franklin. This long lattice bridge had an unusual hipped roof angling down to enclose both roadway and two sidewalks, one of the two such bridge roofs in the Northeast. (The other was at Norridgewock, Maine.)

And then there was the bridge that became a house. When the covered span over Blackwater River at Swett's Mills, New Hampshire, was to be replaced in 1909, the Pearson family had it moved to a hilltop in Webster where it still stands; it serves as a summer home, although it has been rather neglected in recent years.

The railroads, too, had their share of one-of-a-kind bridges in New Hampshire. At Bow Junction, below Concord, was a unique crossing of the Merrimack. Here a covered bridge with gantleted, or overlapping, tracks served both the Boston & Maine Railroad and the trolley cars of the Concord & Manchester Electric Railroad. In Concord proper was another railroad rarity, a long covered lattice overpass across the freight yards. Most unusual of all was a covered railroad bridge on the eastern edge of Keene: it carried rails of the Boston & Maine over the Fitchburg line which in turn crossed a public street.

The Northern Railroad of New Hampshire, a B&M predecessor, ran from Concord to White River Junction, Vermont, and passed through Lebanon, New Hampshire. Here were several crossings of the Mascoma River, including a covered bridge set on a curve. To avoid side sway this bridge was guyed to the shore by a series of heavy rods and turnbuckles. One fine day in 1883 some boys—later described mildly as "mischievous"—loosened the turnbuckles and watched, entranced, as a passenger train set the bridge to vibrating violently. The train had hardly reached the depot beyond when along came a freight, hauled by the road's pride, the heavy engine *Atlantic*.

This time the boys got a real eyeful. The engine had barely nosed into the shaky bridge when the structure began to let go. It broke square in the center and sagged into the river like a tired old sway-backed mare. The *Atlantic* came charging on through, jumped the track and toppled onto her side on the further bank. No railroad men were hurt, but there were some mighty sore youthful backsides in Lebanon that night. The railroad people didn't try to salvage a thing—they just burned up the bridge wreckage and built a temporary trestle of new wood across the Mascoma in eleven hours flat.

As toll charges came to be abandoned on a number of Connecticut River bridges, the owners of those that continued to collect tolls became increasingly unpopular and their practice was called "an interstate holdup." One by one the adjoining towns and the State of New Hampshire bought out the bridge companies, and the little toll houses and familiar gates disappeared.

The last covered toll crossing of the Connecticut was the Cornish-Windsor Bridge, built in 1866, where rates ranged from 2 cents for a foot passenger to 20 cents for a carriage pulled by four horses. Many celebrities, including Presidents Hayes, Theodore Roosevelt and Wilson, have paid to pass under the guillotine-like drop gate and staggered portals of the old Windsor Bridge. It collected its last toll on May 31, 1943. The next day the old bridge was thrown open to non-paying customers with appropriate ceremonies.

In recent years New Hampshire has done a fine job of renovating the old bridge. It's a big structure, 460 feet overall, and is now the longest covered bridge in the United States. It has two clear spans measuring 203 feet, 7 inches and 204 feet, 6 inches, the latter only 5 feet, 6 inches shorter than the world's longest single-span covered bridge at North Blenheim, New York. (The extra overall length is accounted for by the timber lattice work of Windsor Bridge which extends considerably beyond the abutments.)

James F. Tasker of Cornish, a black-bearded and bushy-browed man of iron build, was in charge of erecting the Cornish-Windsor Bridge. With a partner, Bela J. Fletcher of Claremont, New Hampshire, he employed an adaptation of the Town plan, using heavy squared timbers rather than plank to form the web of the lattice. The partners put up other bridges, none of which remains today, on this same timber lattice plan: these include crossings of the Connecticut at Orford-Fairlee and Hanover-Lewiston, and the Pemigewasset River at West Campton, New Hampshire.

The big rivers spanned, Mr. Tasker turned his attention to the lesser streams. For these he designed a truss employing a series of multiple kingposts. Like others before him, Tasker first built a model. It was eight feet long and thirteen inches square and was made of light wood, and it gave little appearance of strength. When Tasker appeared in downtown Claremont one Saturday afternoon with his flimsy-looking model, he was laughed at by the corner loungers. The builder took the little bridge into a nearby hardware store and without a word beckoned the men inside. Then he began piling kegs of nails onto it, pyramid-fashion. When he had stacked up ten kegs, he climbed up and sat on the top one. Since he was a giant weighing over two hundred pounds, the unbelievers' laughter turned to cheers.

It was on this same design that James Tasker built small covered spans in Cornish and Plainfield, New Hampshire, and in Weathers-

RATES OF TOLL.

Every Foot Passenger, - - - - - - - 3
Every head of live Sheep, Hogs or Calves, - - - - $1\frac{1}{2}$
Every head of Horned Cattle, - - - - - - 9
Every Horse, Jack, Mule or Ox, whether led or drove, - 9
Every Horse or Mule and rider, - - - - - $12\frac{1}{2}$
Every two wheel Pleasure Carriage, drawn by one Horse, Jack or Mule, $18\frac{3}{4}$
and SIX CENTS for every additional Horse, Jack or Mule.
Every four wheel Pleasure Carriage, the body whereof is supported by springs or thorough-braces, drawn by one Horse, Jack or Mule, 25
and TWELVE AND A HALF CENTS for every additional Horse, Jack or Mule.
Every Pleasure Wagon, drawn by one Horse, Jack or Mule, 25
and TWELVE AND A HALF CENTS for every additional Horse, Jack or Mule.
Every Stage Wagon, drawn by one Horse, Jack or Mule, $18\frac{3}{4}$
and SIX AND A QUARTER CENTS for every additional 3d or 4th Horse, Jack or Mule.
Every Stage Wagon, drawn by five Horses, Jacks or Mules, 37
and TWENTY-FIVE CENTS for every further additional Horse, Jack or Mule.
Every Freight or Burthen Wagon, drawn by one Horse, Jack, Mule or Ox, $12\frac{1}{2}$
and SIX CENTS for every additional 3d, 4th or 5th Horse, Jack, Mule or Ox.
And for every further additional Horse, Jack, Mule or Ox, TWENTY-FIVE CENTS.
Every Cart or other two wheel Carriage of burthen, drawn by one Horse, Jack, Mule or Ox, $12\frac{1}{2}$
and *SIX AND A QUARTER CENTS* for every additional Horse, Jack, Mule or Ox.
Every Sleigh or Sled, of any description, drawn by one Horse, Jack, Mule or Ox, $12\frac{1}{2}$
and *SIX AND A QUARTER CENTS* for every additional Horse, Jack, Mule or Ox.

Fine of One Dollar.

For any person or persons crossing the Mohawk Bridge on *Horse* Back or in a Carriage or Sleigh of any description, to travel faster than on a walk, or

For any person to cross said Bridge with Horses, Jacks, Mules or Oxen, consisting of more than Ten in one Drove, or to cross with Loaded Carriages or Sleighs drawn by more than two beasts, at a less distance than 30 feet, each from the other, or

For any person or persons with Carriage or Carriages, Sleigh or Sleighs of any description or with any kind of Beasts to take the left hand passage.

The fine Cornish-Windsor crossing was the last covered bridge over the Connecticut to collect tolls. It is now the longest covered bridge in the United States.

field and Windsor, Vermont. He would cut the lumber on his own farm in Cornish, have it sawed and made ready at the mill in Claremont, and then haul it by wagon to the bridge site. There it was a simple matter for a trained gang of men under his direction to frame and set up a bridge in a matter of days, using previously prepared abutments. A couple of specialists were left to side and shingle the structure while Tasker's dayworkers went on to the next job. Here truly was rural construction genius at work. Several of the bridges built in this manner have survived to the present day. Yet it was only by tracing his name that the man who built them could sign a contract: James F. Tasker could neither read nor write.

Farther upstream, the Connecticut is spanned by two more covered bridges. Bedell's Bridge between Haverhill and South Newbury, Vermont, was destroyed by a freak windstorm Sept. 14, 1979, only a month after being completely rebuilt. It stood fast in the flood of 1927, and again in 1936 when the New Hampshire meadows on the east side were turned into a vast lake. Tons of water and huge cakes of ice beat against the bridge without dislodging it. Markings on the west portal showed the height of the water that flowed through the structure in '36; it seems inconceivable today that the flood could ever have reached such a peak.

Mount Orne Bridge at South Lancaster and Columbia Bridge south of Colebrook are products of the present century, but both are now well past sixty years old. The northernmost covered bridge over the Connecticut is completely within New Hampshire, and formerly served a few farms to the west of Pittsburg, up near the source of the river.

Probably the most photographed covered bridge in the Granite State is that at Bath, crossing the Ammonoosuc River. Photographs, paintings and sketches of this bridge turn up regularly on greeting cards, billboards and direct mail advertising from coast to coast. Pictures of it have been palmed off as representing a bridge in New Brunswick and even

Whittier Bridge near West Ossipee.

This bridge at Stark lures many photographers.

Sawyer's Crossing Bridge over the Ashuelot in Swanzey is one of a group to see on an afternoon ride.

one in Switzerland. Bath Bridge shows new stonework in its piers where it was raised some years ago to avoid a bad railroad crossing and clear the adjacent railroad tracks. Its lofty site and sturdy arch construction have served for a century and a quarter to keep Bath Bridge safe from the swift waters of the Ammonoosuc.

Bath's downstream neighbor, the Woodsville Bridge, is one of the oldest in the state. It has two spans of Town lattice design with added arches and a sidewalk. It was built in 1829, only nine years after Ithiel Town received his patent. The bridge is 278 feet long and is reported to have cost but $2400. Contrast this with the price of a "modern" (1937) covered bridge built over the Contoocook between Hancock and Greenfield, New Hampshire. With concrete abutments and a single span of eighty-eight feet, this one had a contract price tag of $77,000. Cost of wood and labor have advanced a bit since great-grandfather's day!

In addition to this one modern bridge, New Hampshire maintains roofed spans at Bartlett, Groveton and Conway in a state of honorable retirement. A bridge at Bagley over the Warner River was abandoned some years ago, and left on its own. It provided a startling example of why a wooden bridge needs a roof. A large patch of roofing was gone near the north end, and the heavy flooring, exposed to the weather, rotted away in *exactly* the same shape as the hole in the roof. Eventually the damage reached the vital trusswork and stringers and by 1965 Bagley Bridge was a span of the past.

Down in the southwest corner of the state the Ashuelot River bridges in the Swanzey-Winchester area make a tight-knit group that can be visited in an afternoon. Here are five big covered spans in close proximity over the main river and a smaller one over its south branch. The bridge at Ashuelot village has an unusual feature in twin sidewalks, and it is painted white both inside and out.

A New Hampshire river with a mouth-fill-

…rgin Bridge near North Sandwich.

Keniston Bridge at Andover.

Famous Flume Bridge in Franconia Notch.

ing name, the Pemigewasset, flows south from Franconia Notch and eventually becomes the Merrimack. Three covered bridges of different types still cross it. First is the Flume Bridge familiar to thousands of visitors who hike or take a bus to the famous gorge. Although often billed as ancient, this example of the Paddleford truss actually was not built until 1886. A few miles downstream the Pemigewasset is crossed by a railroad bridge at Clark's Trading Post, the only Howe-truss span of its type remaining in New England, where it was first fashioned.

Until 1971, when it was destroyed by fire, Woodstock had a big heavy lattice-arch bridge with an impressive single span of 190 feet. Four bridges on this same site were bolstered by a center pier which the river tore at and repeatedly removed (along with successive bridges), so it was decided to make the crossing with a single span. In 1878 a building committee was appointed to oversee the work. The chairman, a portly individual, came to inspect progress one rainy afternoon, shielding himself from the downpour with a large umbrella. Standing too near the edge, he slipped over the bank and splashed into the river. Down the Pemigewasset he spun, bobbing like a cork, banging into rocks and whirling about in eddies. But before the scared chairman had gone too far the bridge crew, racing along the water's edge, managed to rescue him. When he was pulled ashore, although he had lost his hat, the man's head and shoulders were dry: throughout his short but dangerous journey he had continued to clutch the umbrella tight over his head!

The top span is Ashuelot's twin-sidewalked beauty; Woodstock Bridge formerly straddled the Pemigewasset, and Hancock-Greenfield Bridge was the first of the "moderns."

A fine example of the Long truss crosses the Pemigewasset in the Blair section of Campton. Here a settlement with a hotel and railroad station once stood. When Decoration Day ceremonies were held at the old cemetery nearby, the parade of GAR veterans always broke step in crossing the bridge. If it rained on the big day the services were adjourned at the cemetery and reconvened in the bridge.

West of Dover stands Dover Flattop Bridge reaching over the Cocheco River. This Howe truss span, with hardly any pitch to its flat roof, was built about 1870. All its timbers came ready-cut from a sawmill in East Boston, Massachusetts.

Up in Conway was Joel's Bridge, an old landmark and the next-to-last covered bridge in America which bore the traffic of a United States numbered highway. In this case it was U.S. 302, crossing the Saco River between Center Conway and Redstone.

Long ago travelers from Portland on their way north had to jog to the southeast along the Saco and cross the river at Conway. A new road and bridge beyond the Center, it was reasoned, would shorten the distance by about four miles. The proposal was made at

Scenes like this one at the former Melvin Bridge near Winchester were common when Dobbin was king of the road.

Conway's town meeting, and was voted down.

But Joel Eastman, a lawyer-farmer who lived near the site, wanted a new crossing nonetheless; it would give him a direct route to North Conway. Together with two neighboring landowners, he laid out the new road anyway. Then he called Master Builder Peter Paddleford over from Littleton and hired him to throw a fine new covered bridge across the Saco. Paddleford finished the job in 1846.

Conway people naturally expected to see a toll barrier at Joel's Bridge, for the four-mile shortcut was much traveled even though the townsfolk hadn't been willing to build it themselves. Joel chuckled and bided his time. Then, after a year, he announced that the town of Conway had accepted his bridge by making constant use of it. He showed the uneasy selectmen that under the laws of the state they must pay for what they were using. The town paid for road and bridge.

Joel's Bridge was repaired in 1934, with new splices and ship's-knee braces of tamarack, installed under the direction of Frank Broughton, the last of Conway's old-time covered bridge builders. The trusses were structurally sound, and the state later gave the old bridge still another face-lifting—a coat of red and white paint on the truck-battered portals. What time, the river and heavy traffic could not do to Peter Paddleford's wooden fabric was accomplished by arson in celebration (?) of the Fourth of July, 1975.

The Paddleford Truss.

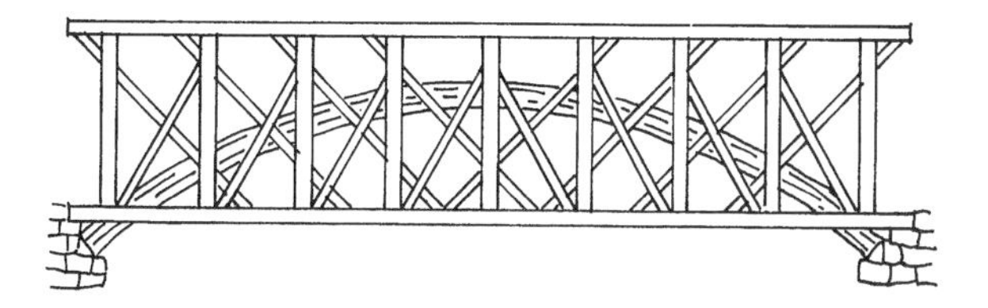

VERMONT:
Collectors' Paradise

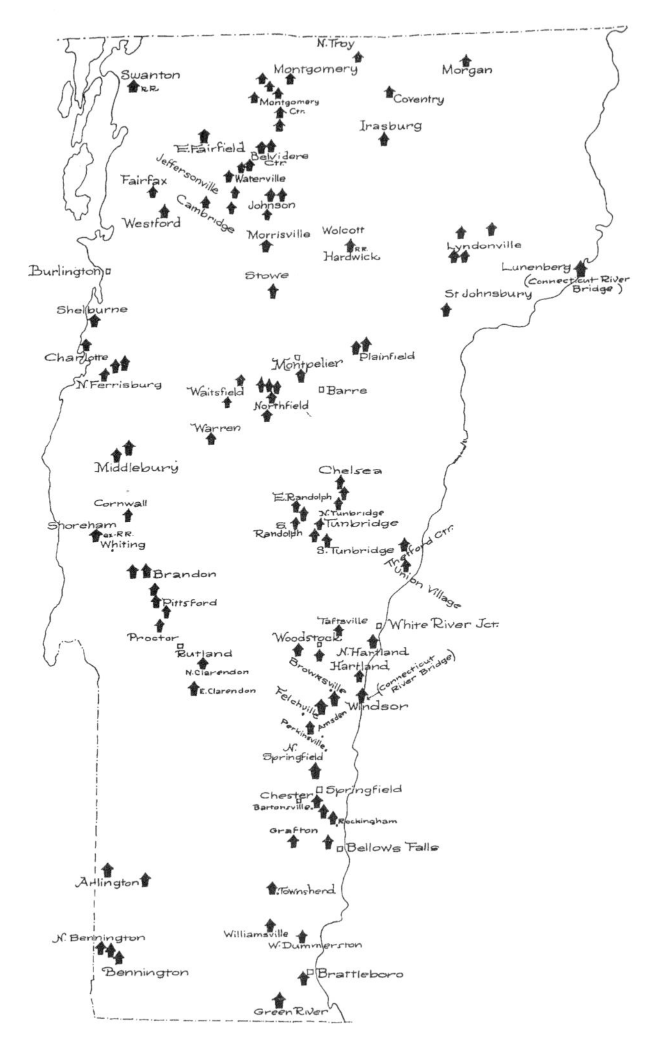

VERMONT IS JUSTLY FAMOUS for her covered bridges. No other state has built and still possesses so many of the old timbered crossings in so small an area. All over the state the little brooks are swift and rocky, the big rivers run broad and deep and roads cross and recross valleys of all sizes: Vermont terrain and covered bridges seem made for each other. Wood was long the natural choice for bridges, with trees growing in abundance on the Green Mountains, and to get it was often just a matter of climbing up the hill behind the barn to fetch down some that had been cut and left to season.

A tabulation of all the covered bridges erected in Vermont will never be entirely complete, but there are records of well over half a thousand that once were dotted thickly over this area of only 9528 square miles. Little wonder that many people think first of Vermont when covered bridges are mentioned.

The state's pioneer builders of covered bridges were probably journeymen carpenter-contractors. They used the arch plan of Theodore Burr, which they had seen over in York State, and sometimes they added improvements of their own. A product of this early period is the Pulp Mill Bridge still standing over Otter Creek north of Middlebury which may date from as long ago as 1820, which would make it the oldest covered bridge in the United States.

But the earliest authenticated building date for a Vermont covered bridge is 1824. In that year two brothers named Keyes built the long single-span, arch-truss toll bridge across the Mississquoi River at Highgate Falls. Lyman Burgess and Roderick Hill built a similar arch-truss bridge over the Lamoille River at Milton, and Sylvanus Baldwin was responsible for Burr-type trusses across the Winooski. These set an engineering precedent, one followed for many years thereafter for bridging the wider rivers in the northern part of the state.

Vermont's most famous covered bridge builder, whose work might be called typical of the middle 1800's, was Nicholas Montgomery Powers,* born in 1817 on a farm southwest of Pittsford in Rutland County.

Powers served his bridge building apprenticeship under Abraham Owen of Pittsford, who favored Town lattice construction. Before he was twenty-one young Powers had convinced the Pittsford selectmen that he was capable of putting a covered bridge over Furnace Brook at Pittsford Mills. His father, Richard Powers, put up bond and agreed to make good any timbers the "boy" might spoil. None was spoiled (and the bridge was strong enough 96 years later to support a 20-ton steam roller operated by state roadbuilders who came to replace the span with a concrete structure). Naturally the Pittsford Mills job

*Relatives and descendents have always referred to this venerable bridge builder as "Nicholas" or "old Nick." More recent research, however, discloses that his name, as evidenced by his tombstone, was actually "Nichols." [R.S.A.]

made young Powers' reputation, and for forty years he was seldom without a bridge contract.

Nick Powers had plenty of work close to home the first dozen years or so, bridging Otter Creek, East Creek and Cold River in the Rutland area. He favored spruce and hemlock for his bridges and used to say that a spruce stick was as strong as an equal weight of the brittle iron of his day. In 1849 one of his jobs was to build a bridge over East Creek on the Chittenden road. The next Spring freshet changed the course of the stream, leaving the bridge high and dry. Powers sold the Rutland selectmen on the idea of another structure to span the new channel instead of moving the old span. His salesmanship resulted in the Twin Bridges which took care of all the high water that East Creek could throw at them until the Pittsford Dam broke in 1947. One of the wrecked twins was later converted into a storage barn south of the former crossing.

In the 1850's Powers began working farther away from home, leaving hired men to run his good farm in Clarendon during his frequent absences. Down on the railroad line from Boston he helped bridge the Connecticut River at Bellows Falls. This put him in touch with other bridge men, real college-trained engineers and full-time professional railroad contractors. Nick wasn't much on "book-learnin' " but he was a whiz at the practical business of joining timbers. He began to experiment with models, dreaming up new designs and testing them for strength.

In 1854 he was called to North Blenheim, New York, to bridge Schoharie Creek. Here he let his imagination have full play and built a covered bridge the like of which had never been attempted before—nor has it since. This was a 228-foot two-lane bridge with three large trusses based on Long's design. It is today the longest single-span covered bridge existing in the world, and from an engineering standpoint is a real oddity among bridges. We'll describe it later in the discussion of New York bridges.

When the Blenheim bridge was nearing

Vermont's first known covered bridge (1824) used Burr's arch-truss over the Mississquoi at Highgate Falls.

NICHOLAS M. POWERS

Truss detail in Old Blenheim.

completion in 1855 there was some question as to whether the mighty span would stand at all. Local sidewalk superintendents clucked tongues and called it Powers' Folly. "Much too heavy!" they declared. "It'll fall of its own weight!"

To show his faith in his masterpiece Powers clambered up to the ridgepole. "If she goes, I'll go with her," he said. Then he called down to his crew to knock away the falsework. The crowd drew its breath as the last supporting timber splashed into the river and the bridge creaked and settled—just a fraction of an inch. Up top, Nick Powers sat grinning: he had succeeded with his experiment.

The farthest from Vermont that Nicholas Powers got was the mouth of the Susquehanna River at Perryville, Maryland. There George A. Parker, an engineer with whom Powers had worked at Bellows Falls, was charged with building an enormous railroad bridge. Nick was first employed as a boss-carpenter, one of many. Early in 1866 a tornado demolished all but one span of the nearly finished bridge and the hired designer was thereupon fired. Two of the railroad's best college-educated draftsmen worked for a fortnight on new plans without much progress. In desperation Parker came to Powers and asked him how long it would take to design a new bridge that would stand up. Looking

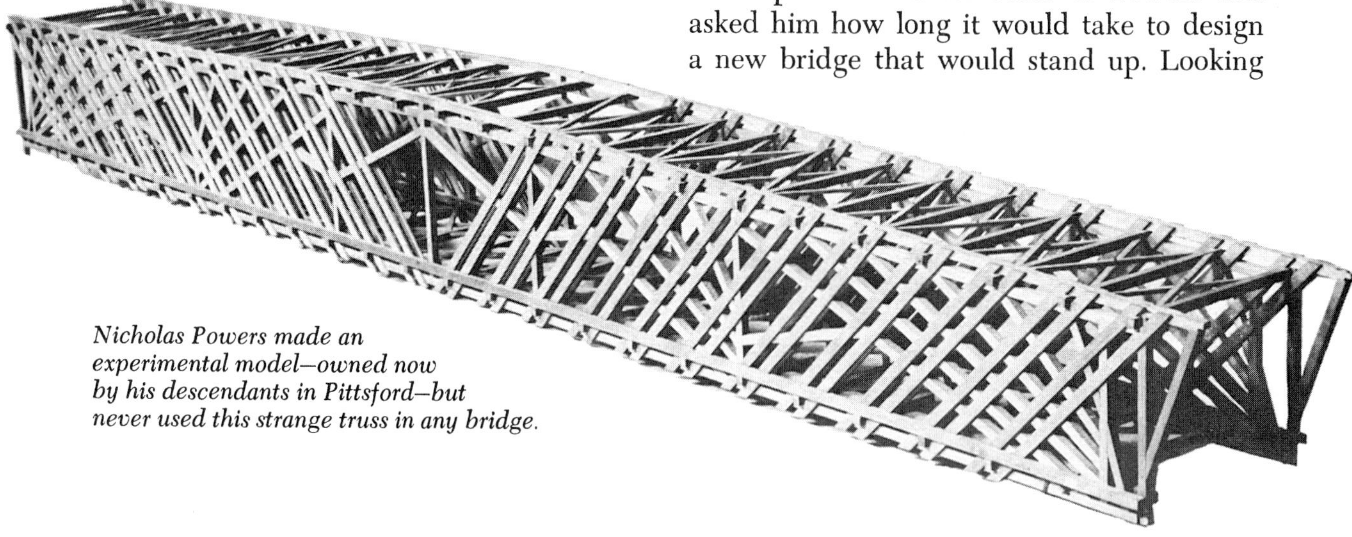

Nicholas Powers made an experimental model—owned now by his descendants in Pittsford—but never used this strange truss in any bridge.

Powers used his favorite lattice truss in little Cooley Bridge over Furnace Brook near Pittsford.

at his watch, which then said a quarter of ten, Nick replied: "I'll give it to you after I've had my lunch."

Sure enough, when Parker came back the Vermonter had covered the sides of a big timber block with calculations and drawings of what appeared to be a mighty promising bridge. Then and there the chief engineer gave Powers the contract. Nick was to get $7 a day and his choice of fifty men to work with. He ended up bossing four hundred and finished the bridge ahead of time to claim a $500 bonus. And to confound the engineers further Powers had his sixteen-year-old son Charles design the drawspan.

Back home in Clarendon Mrs. Powers didn't think much of her husband's being away for so long, and wrote him as much. Nick answered:

"If you could see this work going ahead and the place I hold, I think you would tell me to stay 'till the job was done. I am treated with more respect here in one day than I would be in Clarendon in one year. . . . When I get done on this job you may reckon on my staying to home after that."

Nicholas Powers kept his promise. He left Maryland as soon as the great Susequehanna bridge was done, not even waiting to celebrate the crossing of the first train. Although he railroaded a bit as a division engineer on the Bennington & Rutland Railroad, he never again left Vermont. Nor did he try anything more in the way of spectacular design; his later bridges were all on the old time-tested Town lattice plan which he had worked with as a boy.

Some say that members of the Powers family were "so strict they wouldn't crack a nut on the Sabbath," but this does not give a true picture of the Vermont bridge builder. As a lad Nick was the kind who would feign illness so that he could stay home from Sunday school and sample the applejack in the root cellar. An inveterate practical joker, one of his favorite tricks in later years was to place some strong Limburger cheese on his bald pate, clap his hat tight down over it, and go downtown in Rutland to mingle with a crowd. When people's noses began to wrinkle old Nick would begin to sniff with the rest and cast disapproving glances about him.

As time went on the stories of Nick Powers' work grew to Paul Bunyan-like proportions, and some must be classed as folklore. Writers have glibly credited him not only with "all the lattice bridges in Vermont," but with "most of the covered bridges in New England and New

York." If Powers had built all the bridges attributed to him, he would have had to be working day *and* night for forty years. But the records of the ones he did build make him a man well worthy of the legends.

At least three of the covered bridges that Powers oversaw or worked on himself still stand in Rutland County. In Pittsford he helped Abraham Owen build both the Mead and Gorham bridges which cross Otter Creek. Over nearby Furnace Brook stands Powers' own Cooley Bridge, a pint-sized specimen that has been compared with a stranded prairie schooner because of its deeply overhung portals. The master builder's last known job was the Brown Bridge which spans Cold River deep in the woods east of Clarendon. Powers built this one for the Town of Shrewsbury in 1880, concluding more than four decades of covered bridge building.

The Burr truss design used so extensively in northern Vermont was also the style for well over a dozen double-barreled covered bridges situated for the most part over the rivers flowing into Lake Champlain. Bridges separated for two-way traffic were also used by the railroads; the big one leading into Bellows Falls from New Hampshire, and a smaller one on the Delaware & Hudson Railroad over East Creek in Rutland were good examples of the type. Down in West Hartford a covered bridge of the divided-lane type figured in an odd and tragic accident. John Steele, a young merchant of West Hartford, set out with some village boys one Fall night in 1848 to "inspect" some of the neighbors' melon patches. Apparently the neighbors were snug in their beds, for the boys gorged themselves and took along some spoils for further consumption. As they trudged back to town they passed the big double-barreled covered bridge across White River. A late driver coming up from White River Junction scared them, and to escape possible recognition the boys turned and dashed into the span's protecting shadows. Unfortunately the town was then in the process of renewing the flooring, and during the day the boards had been removed from the south side. Forgetful of this fact, young Steele clambered over the dividing truss between the two passageways and then over an obstruction put up to confine travel to one lane. Leaping off the last barrier with no thought but of flight, he hurtled to the rocks fifteen feet below. Steele was severely injured and never fully recovered from his fall. Surprisingly, he received $1500 in compensation from the town even though it had dutifully and carefully barricaded its bridge.

There were other tragic happenings in Vermont covered bridges. In Wallingford late one night a man hanged himself from a crossbeam. His corpse was cut down by a passing traveler—who always by-passed that particular bridge on future trips.

South of Rutland, Avery Billings was set upon by masked men as he drove through the bridge between his farm and the city. One leaped from the shadows to grab the bridle of the horse while another dropped from the cross bracing into the wagon. The horse plunged and bucked as Billings lashed out with his whip at one assailant. The other lost his grip on the dashboard and was bowled out onto the bridge floor. The farmer escaped.

Isaac Kelly was not so lucky. The Kellys never did get a bridge of their own directly across Otter Creek to Rutland: they had to cross either by Dorr's or Billings' bridges. Isaac started out for Rutland one night to buy a railroad ticket for his mother; he was never seen alive again. Days later a trail of bloodstains was found, leading from the interior of Billings' Bridge to a clump of brush nearby. There lay Kelly's body; he had been beaten and robbed. The murderers were never found.

Though it seems a most unlikely place, St. Albans was the scene of a dashing daylight raid by Southerners late in the Civil War. The idea was to get some money for the ailing Confederacy and to throw a scare into the

The artist flattened terrain near Black Creek Bridge, making the St. Albans raiders' getaway extra dramatic.

complacent Northern cities and states. Down into Vermont from Canada came Lieutenant Bennett H. Young and twenty-one daring Confederate soldiers traveling quietly and wearing civilian clothes. At three o'clock on the afternoon of October 19, 1864, they donned their uniforms and simultaneously entered the three banks of St. Albans, killed one man and wounded others, and dashed off with $208,000. Ten miles out of town the raiders threw phosphorous into the big double-lane covered bridge over Black Creek at Sheldon, and set it on fire. The people in the village managed to put the fire out and save the bridge. The Confederates escaped across the border into Canada.

Sheldon Bridge finally did burn in the 1920's. And soon thereafter another big bridge at Jeffersonville met the same fate. In the Fall of 1929 a wagonload of hay in a nearby meadow caught fire. The frightened horses bolted for the road, dragging the blazing hay behind. Turning into the bridge across the Lamoille, the doomed animals jammed the fiery load in the passageway and everything went up in flames.

Intentional burning was the fate of the Casey Bridge over Mad River north of Moretown. Fitted with doors at each end, the old lattice bridge served for several years as a storehouse for town road supplies and machinery after the wooden span was by-passed by a new steel structure. At length it developed a broken lower chord and a terrific sag and was no longer useful. Sprayed gasoline and a match finished it off.

Vermont economy has often been evident in covered bridge building as well as maintenance. A 75-foot bridge across the Ompompanoosuc River at Post Mills cost $543.25 in 1878. Four years later a little 30-footer at West Dover made a record for low cost in covered bridges. John B. Davis, a local mill owner, built this bridge over the North Branch of the Deerfield River for just $149. And it stood for fifty-six years.

Henry Campbell of Philadelphia originally came to Vermont to build railroads and railroad bridges, eventually settling down in Burlington to take an occasional local job. In 1860 he contracted with the selectmen of Burlington and Colchester to put up what was known in later years as the Heineburg Bridge, spanning the Winooski River on the flats north of the city. Mr. Campbell was mighty proud of the bridge and assured the selectmen that it would "stand until the foundations of all things shall be broken up." The builder was

The Passumpsic rages through St. Johnsbury, slamming Hastings Street bridge against Arlington Bridge in the disastrous flood of 1927.

asking a lot of a wooden structure, but Heineburg Bridge *did* become one of the very few covered bridges over the Winooski to survive the great flood of November, 1927.

If it had not been for the '27 flood, Vermont would have many more covered bridges than are standing today. The exact number has never been compiled, but it is estimated that nearly two hundred covered wooden spans in the Green Mountain State were destroyed within a space of forty-eight hours. The rain seemed to concentrate for days on Vermont; water in rivers and streams crept higher and higher, above the high water mark of 1913, above that of '69, higher than any living man could remember.

Old bridges that had stood as landmarks for seventy and eighty years were torn from their foundations and sent hurtling downstream. At St. Johnsbury the Hastings Street Bridge, with water over its flooring, withstood the tide for hours, then sailed off like a frigate down the Passumpsic River. It crashed into the Arlington Bridge and, locked in splintered embrace, both went crashing over the falls beyond. Dorr's Bridge in Rutland took out the Ripley Bridge and after it a giant iron railroad bridge supposedly far too high above Otter Creek to be threatened. North Enosburg Bridge came floating down the Mississquoi and ended up in the yard of a house at Enosburg Falls. Above Bridgewater the swollen Ottauquechee cleared the valley of nearly a dozen covered bridges. All over Vermont the same thing happened.

Sometimes the bridges stood firm. At Riverton a tree, borne by the rampaging Dog River, was driven through both walls of a stalwart covered span. Knocking off the sideboards saved several covered bridges, enabling the water to flow through them unimpeded.

Residents had a hard time with the big village bridge at Cambridge. A hay-filled barn came floating down the Lamoille, pushed from side to side by the buffeting current. When it grounded temporarily above Cambridge, two men rowed out to it in a boat and set the hay afire, hoping to destroy the barn and so save the bridge. The barn had hardly started burning when it pulled loose and once more set off downstream—it was now a double danger. Cambridge folk watched breathlessly. Seemingly at the last moment the blazing barn snagged again, and burned itself out on the flats. Big Cambridge Bridge was saved.

At St. Johnsbury the high-portaled Maine Central Railroad bridge was set on fire deliberately so that it would not tear away and take out the steel bridge across the Passumpsic River just below. This effort, too, was successful, and when the waters receded the steel bridge remained the only link between the two parts of town.

There wasn't much to salvage from the modern bridges after the flood of '27. Recovered steel could be sold as scrap, to be sure, but slabs of concrete were a dead loss. The wooden wreckage of the covered bridges, however, provided thousands of feet of timber for temporary trestles and foot bridges during the recovery period. In several cases a bridge was deposited almost intact in some field miles downstream. A bridge over the Winooski at Waterbury, left in this manner, was rebuilt

Rare picture taken in 1887 shows construction crew at the span over the Lamoille at Cambridge Junction. George Washington Holmes (second from right) *built the bridge, which still stands today.*

into a crossing of Little River, west of town. Hammond Bridge in Pittsford floated off down Otter Creek, but was rescued and towed back to its abutments before the water went down. Over in Stockbridge a small span over the Tweed River had a similar adventure, catching on a big willow tree and later being restored to its former site. In 1942 Mrs. Elsie Bindrum wrote a fine children's book, *The Little Covered Bridge,* based on this incident.

Vermont today still has covered bridges of many types in a great variety of settings. They are found in all but one county: the low-lying islands which make up Grand Isle County—where the streams are tiny—apparently never had any.

Bennington County, settled early in Vermont's history, has fine old lattice bridges over the streams which drain westward into New York State. The three over the Walloomsac River northwest of Bennington are all horizontally sided in the old-time manner, and are painted red with white trim. All three were handsomely restored in 1952–53. The Town of Bennington appropriated $35,000 for the work and placed Chauncey Donaldson in charge of it. Henry Bridge, the most westerly span, had been reinforced in Civil War days with ten thicknesses of plank and timber lattices. This extra bracing had been thought necessary when iron ore was being mined on the hill beyond the bridge and hauled to North Bennington for processing. Mr. Donaldson had this needless dead weight of timber removed and today Henry Bridge, at one time called "the strongest covered bridge in Vermont," has once again become a single-web Town lattice structure.

A story about the Arlington Green Bridge over the Batten Kill furnishes further proof of the inherent strength of lattice spans. Shortly after the bridge was built a freshet undermined the abutments and toppled the bridge over on its side. In this shape it was used for several months before being righted, over one hundred years ago.

To the north of East Arlington is the well-known Chiselville Bridge built high above the Roaring Branch of the Batten Kill after previous bridges at a lower level had been destroyed by floods. The danger now is from cold wintry blasts sweeping down from the mountains, and long wire cables guy the Chiselville Bridge against wind pressure.

Addison County has only five scattered covered bridges. At the Marble Ledge northeast of Middlebury stands the highest covered span in Vermont, a short lattice bridge 41 feet above the Muddy Branch of the New Haven River. A two-town bridge spans Otter Creek in the Great Cedar Swamp between Salisbury and Cornwall, and the North Ferrisburg former village bridge is now to be found alongside U.S. 7. At East Shoreham the lonely old Rutland Railroad Bridge over the Lemon Fair River has been shorn of its tracks to serve as an access to a state fishing preserve.

The Pulp Mill Bridge north of Middlebury, referred to previously, completes the roster of Addison County bridges. It is one of eight existing double-barreled covered spans in America and has been a notable Otter Creek landmark for well over one hundred and fifty years. Originally built as a clear span of 179 feet, the big bridge was long ago strengthened by auxiliary piers. The main single-timber arches have been reinforced by heavier ones of laminated plank. At night old Pulp Mill, lit only by cowebbed, dirt-encrusted street lamps, presents a spooky passage.

Chittenden County shows the influence in this part of Vermont of the arch-truss bridge. There are four small examples in the town of Charlotte and one at Westford. Most notable and accessible is, of course, the big double-barreled Burr-type arch bridge beside U.S. 7 at Shelburne. It serves as an entrance to Shelburne Museum, a re-created New England village.

Bridge at Shelburne Museum.

A man named Farewell Wetherby built this bridge across the Lamoille River at Cambridge, Vermont, in 1845. As we have seen, it was nearly destroyed in the flood of '27. When it was due to be replaced in 1950, no one dreamed that such a massive structure—it is 168 feet long—could ever be salvaged. But Mrs. J. Watson Webb, moving spirit of the Shelburne Museum, had other ideas. She wanted a good example of a covered bridge at Shelburne, and the Cambridge Bridge was undoubtedly one of the finest specimens in Vermont.

Having been given the bridge by the highway department with the understanding that

Car versus *bridge: the entrance to Dorr's crossing at Rutland proved too much for this Model T. Many sound bridges in the Northeast were considered obsolete after automobiles came along.*

she would remove it, Mrs. Webb engaged Professor Reginald V. Milbank of the University of Vermont, a civil engineer, and W. B. Hill and Company of Tilton, New Hampshire, to undertake moving the bridge to Shelburne. They took the bridge apart carefully, trucked it forty-five miles to its new site, and there re-erected it. A small pond was excavated under the bridge to give it a natural setting. The bridge stands today as both an attractive approach to the museum and an outstanding example of covered bridge preservation.

When it moved to Shelburne it left a smaller twin behind in Cambridge. This twin was jacked up on new abutments and the channel of the Seymour River altered to flow beneath it; today it serves as an access to Earl Gates' farm meadows. George Washington Holmes, its builder, would hardly recognize it in its present location. But he would have no such difficulty in identifying another of his spans, for the bridge across the Lamoille River at Cambridge Junction looks today almost as it did when it was built about seventy years ago. Located close to the railroad station, it was a favorite place to park a horse and rig while waiting for the train. A large board sign used to prohibit this practice, but unthinking souvenir hunters have removed it.

Other Lamoille County covered bridges include a group of five in Waterville and Belvidere, spanning the North Branch of the Lamoille; three in Johnson, and back-road bridges in Stowe and Morristown.

In Hyde Park a little span once perched over a rocky glen at the crossroads hamlet of Garfield. Garfield is apparently a tall-tale center. Willard K. Sanders, a Morrisville historian, has more than once heard a hair-raising yarn to the effect that some fifty years ago the entire community used to snow the bridge and then go out to slide of a Winter's evening. They used a traverse or bobsled some 45 feet long and 8 feet wide, which could seat forty people. Down Davis Hill it would plunge, two men to steer and a third, the navigator, to aim it so as to hit the narrow passage of the bridge with only inches to spare. Mr. Sanders also reports he has been told that the Garfield Bridge is "over two hundred years old." He says that if he could believe the span was erected back when the country was inhabited only by Indians and mosquitoes, then he could also believe the story of the giant traverse threading the needle's eye of the bridge.

West Hill Bridge near Montgomery.

The former Levis Bridge in Montgomery was built by Sheldon and Savannah Jewett on the lattice plan.

Except for the small bridges at Fairfax and East Fairfield, Franklin County's covered bridges are confined to the town of Montgomery. Here are seven Town lattice bridges in close proximity, all built by the Jewett brothers, Sheldon and Savannah. For more than thirty years the Jewetts were called on to build Montgomery's bridges, preparing their own timber and hauling it to the bridge sites. One of the Jewett Bridges, near where their mill once stood far up West Hill Brook, is now privately owned by a resort operator in Highgate Springs. In a beautiful woods-and-water setting, the short hike required for a visit to it is most rewarding.

Still standing in Orleans County are the two Paddleford truss bridges built by John D. Colton of Irasburg. And over the Missisquoi south of North Troy is an odd, buttressed bridge, with lattices joined at the intersections by a single pin—something of a rarity.

Lyndon, in Caledonia County, has four covered bridges, rather ungainly structures with high wide roofs; and there is a lone specimen of the same type at the ghost town of South Danville.

Herman Haupt's Truss.

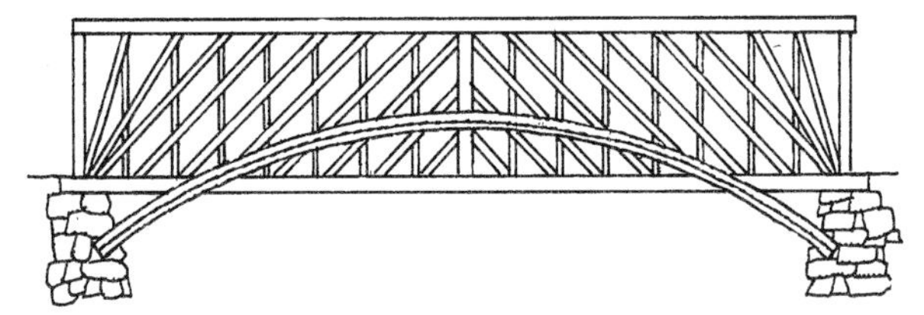

Discounting New Hampshire's Connecticut River bridges, Essex County can boast of only a few feet of two covered bridges, the portions that are officially Vermont's of New Hampshire's Connecticut River crossings in Lemington and Lunenburg.

Orange County still clings to its covered bridges in Chelsea, Randolph, Tunbridge and Thetford. The one at Thetford Center is of interest because it is the only Haupt truss covered bridge in the Northeast. Resembling a combination of the Town lattice and multiple kingpost, the truss was invented in 1839 by Herman Haupt of Gettysburg, Pennsylvania. More famed as Civil War general and builder of the Hoosac Tunnel than as a truss designer, his textbooks on bridge construction nevertheless went into many editions and were widely read. It is presumed that the Haupt patent truss design found its way to Thetford in this way. In any case, the design was faithfully executed in the bridge across the Ompompanoosuc River, and it has neat auxiliary arches besides.

North of Royalton, the First Branch of White River was once crossed by fifteen covered bridges in a space of only eleven miles. Tunbridge had nine of them—it still has five today—all built on the multiple kingpost plan. In the old days the stage that ran between South Royalton and Chelsea passed through three of them. The portals of these bridges

Comstock Bridge is a good example of the Jewett brothers' building jobs around Montgomery.

bore the usual white boards reading: "$5.00 Fine for Driving Faster than a Walk on this Bridge, per order of the Selectmen."

C. F. Peters owned and operated the stagecoach in this part of the county. He drove four white horses at a spanking clip up and down the valley and oftentimes he would run the bridges at high speed. One day he was hauled in for this by the Tunbridge constable and fined the usual sum. With a lordly gesture he gave the magistrate a ten dollar bill.

"Keep the change," he boomed, "I'll be back tomorrow and run your damn bridge again!"

At Waitsfield, in Washington County, a side road leads over Old Arch Bridge, which dates from 1833, and north from remote Waitsfield Common is a little span crossing Pine Brook. At Northfield Falls are three covered bridges in a short row: a big specimen over Dog River and two smaller ones over Cox Brook.

Hectorville Bridge on the Trout River, Montgomery.

Two covered bridges still span the upper Winooski. The Orton Farm bridge, a private one close to U.S. Route 2, is the last remaining covered span built by Herman Townsend of Marshfield. Another Townsend-built bridge, on an unfrequented town road leading to just one farm near Marshfield, once served as a gathering place for young bandsmen. The boys rigged up the bridge by laying plank across the lateral bracing to make a kind of second floor. Lanterns by which they could read the music hung on rafter nails, and sap tubs were used as seats. Six or eight would-be musicians could while away many an evening hour in this lonely but cozy retreat, free from interruption and with nobody to mind their discords. One traveling thread salesman got the scare of his life when he passed the bridge on a gloomy night. His friends wouldn't believe it, but he swore to ghostly revels, with a strange light and music coming out of an empty bridge you could see clear through!

In addition to the long Cornish-Windsor Bridge which touches its Connecticut River shore, Windsor County has some unique covered structures. Though none of James Tasker's little multiple kingpost truss bridges still stand in Weathersfield, near Brownsville are two rare spans whose truss-work consists only of small laminated arches. The big Taftsville Bridge over the Ottauquechee River is a strange hybrid of bridge trusses. But the most unusual one in the country is the Lincoln

Bridge near West Woodstock: the only known wooden example left in America which shows the principle of bridge construction devised by Thomas Willis Pratt.

Willis Pratt was the son of a prosperous Boston architect. A child prodigy, at the age of twelve he was making the difficult working drawings for a house his father was to build, and at fifteen was offered a teaching position at the technical school in Troy, New York, which is now Rensselaer Polytechnic Institute.

With his engineering education, young Pratt, like so many others, found a fertile field for his talents in the railroad-building ventures of the 1830's. Barely of legal age, Willis served as aide, division engineer and then superintendent of the Norwich & Worcester Railroad in Connecticut. While living in Norwich in 1844 he invented and patented the bridge plan still known today as the Pratt truss. This was just the opposite of Howe's plan, using upright vertical posts of wood, with braces and counter braces of iron.

The engineer received little or no money for his patented invention, for, in wood, it was awkward to erect and difficult to adjust. As years went by, however, it became a standard railroad bridge when built of wrought iron and later of steel. By that time the patent had expired. While working for the Eastern Railroad after the Civil War, Pratt invented a new type of double-web lattice bridge which was an improvement on Ithiel Town's old plan. This type of Town-Pratt bridge was built on many rail lines in northern New England.

Willis Pratt was a heavy-set, bearded man who wore rimless eyeglasses. He was noted for his wonderful memory. Out of his head

Lincoln Bridge (Pratt type truss) at Woodstock.

Former span over Taft Brook in Westfield.

he could write down all the formulas and proofs for any engineering device he had ever used, a handy accomplishment at a remote bridge site far from books and drafting tables. Though he never received proper recognition, he certainly fulfilled his promise, and his monuments today are the hundreds of giant steel Pratt-type railway and highway bridges all across the nation.

Best-known in Windham County is the Creamery Bridge, seen by thousands who cross the state on the Molly Stark Trail and pass it coming into Brattleboro. The valleys of the Deerfield, Williams, Saxton's and West Rivers have them too, ranging in size from the rebuilt 42-footer from West Townshend at the Victorian Village in Rockingham, to the 280-foot, two-span structure which stretches across West River above West Dummerston.

Caleb Lamson took a disastrous tumble during the building of the West Dummerston Bridge. Caleb, only twenty-two, was contractor for the big crossing. He was standing on top of the truss halfway across the river when the scaffolding suddenly collapsed. Wisely staying put until he saw which way the truss was going, Lamson jumped in the other direction, down into shallow, rocky water. Although he survived his back was permanently injured. One of his workmen, intent on the boss' predicament, failed to see his own danger and was killed by a falling timber. Lamson rebuilt immediately, and his massive structure is today the longest covered highway

A workman lost his life during construction of West Dummerston Bridge, now the state's longest.

bridge in Vermont.

Only four feet shorter is Scott Bridge, over West River in Townshend. In 1955 the town deeded this bridge to the Vermont Historic Sites Commission. It is the first of a projected "sampling" of Vermont covered bridges which will eventually be acquired and maintained by the commission in all parts of the state.

The *Brattleboro Daily Reformer* tells of an incident which occurred in Scott Bridge. At one time Seth Allen of West Townshend was hired to reshingle the long structure. Allen was good at his work and nimble enough to need little but a ladder in carrying it out, but for some obscure reason he enjoyed working by moonlight. Late one night a man who lived over on the west hill drove home from town by way of the quiet darkness of Scott Bridge. Halfway across, a terrific pounding began over his head. The horse left no time for investigation and took off for the tall timber as though the devil were at his heels. Mr. Allen was subsequently required to please do his bridge shingling at a more orthodox time.

The covered bridge at Green River in Guilford serves a triple purpose. Besides carrying travelers across the stream it is also the storage rack for the town derrick, and along its dim interior are the rural mailboxes for the neighborhood.

As related in Walter Hard, Sr.'s *Vermont Vintage,* the scene of one of the best covered bridge stories was Newfane, where two covered bridges survive today. Time was when the road up West River crossed and recrossed the stream and its branches by means of covered bridges. An old Civil War pensioner from Newfane celebrated properly in Brattleboro one Saturday night. Some time in the early Sunday hours a farmer below town was awakened by a racket in his buggy shed, which stood out parallel to the road. Pulling on pants and boots, the farmer took his lantern and went to investigate. There was a horse and wagon, jammed tight into the shed. The pensioner sat disconsolate, his legs dangling over the tailgate, and his eyes blinking at the light thrust in his face.

"All I want to know," he said, "is who in 'ell boarded up the end of thish bridge!"

Like several spans that were once main village crossings, Mill Brook Bridge in Fairfax is little used today.

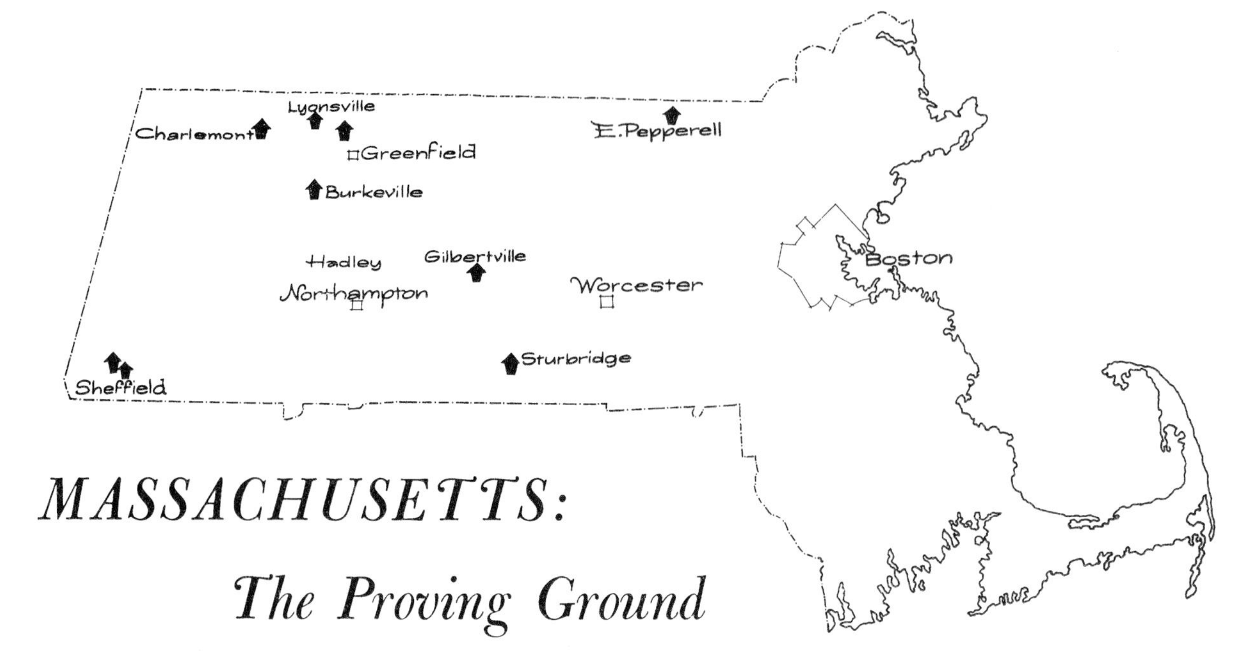

MASSACHUSETTS:

The Proving Ground

WHEN YOU CONSIDER that Timothy Palmer, builder of America's first known covered bridge, was a native of Massachusetts and first tried out his plans in the northeast corner of the commonwealth, it will be understood why the Bay State has been called "The cradle of covered bridge building in America."

Palmer hailed from Rowley. Thinking big, he didn't waste time with bridges over small streams: he started right in to span the wide Merrimack. He built great wooden arches across the river at Lawrence, Haverhill, Rocks Village and Deer Island above Newburyport. For more than thirteen years he left his heavy timbers and their vulnerable joints exposed to the weather. In 1805 he finished his huge three-span Permanent Bridge over the Schuylkill, and that year a Philadelphia judge convinced him of the worth of enclosing his spans, thus giving the first known covered bridge to Pennsylvania and robbing Massachusetts of that honor.

Palmer's Haverhill Bridge over the Merrimack.

Later, around 1810, the shorter of Palmer's two Deer Island bridges was weatherboarded and roofed. Since its "innards" had been built in 1792, it can be said that this structure was *made into* the first and oldest covered bridge in the United States. It was a humpbacked bridge with a 113-foot span, built with a double lane which rose up and over the Amesbury channel of the Merrimack River. From a distance it resembled a half-squeezed accordion.

At Haverhill was another Palmer product which was remodeled and given a roof in 1825. For nearly fifty years the tightly enclosed structure was a picturesque landmark of the town, with the graceful lines of its siding painted a gleaming white. The bridge's outside walkway made a good promenade and was favored over the inky black interior except in the wintertime.

A third Palmer bridge at Rocks Village halfway between the first two Merrimack spans lasted less than twenty-five years before it was taken out by ice. Its successor, built on Mr. Town's lattice plan in 1828, had a long career. Originally stretching 802 feet, it had four spans and a draw. Over the years span after span was replaced by iron until only one wooden section remained. In 1916 this, too, gave way to iron.

Records of the first bridge at Lowell, farther

up the Merrimack, show that liquid refreshment played a big part in early bridge building. The purchasing agent for the bridge company was empowered to buy "a ton of iron (probably bolts and washers) and two barrels of New England rum." The rum was exhausted before the iron was, and two months later another barrel had to be bought—plus half a barrel of the West Indian variety for the bridge proprietors, who presumably had finer tastes than the workmen. To top it off, everyone who crossed the bridge on the opening day in 1792 was treated to a mug of flip or toddy!

The Bay State's other big river, the Connecticut, divides the center of the commonwealth, and was once spanned by eight covered bridges. At Northfield was a combination bridge with railroad tracks placed on top and the roadway inside. When a train went over even the best of teams would rear up on their hind legs and take off for the safety of the far shore. It was black as a pocket on the bridge at night, and if you crossed after nine o'clock you were required to carry your own lantern.

Not far downstream was a similar bridge between Greenfield and Montague City, an 863-foot Howe truss on high piers. This one, too, was tightly boarded and very dark, and here a horse actually died of fright when a Fitchburg train thundered across overhead. Railroad traffic was discontinued in 1921, but by that time horses were scarcer. The great ice flood of March, 1936, swept the whole bridge down the rampaging Connecticut. At the time of its destruction Montague City Bridge was the longest covered bridge in the United States.

The Sunderland-Deerfield crossing was an unlucky spot. Six successive covered bridges over the Connecticut here were washed away and the seventh was laid low by a windstorm. A regular tornado on a June afternoon in 1877 did in the third covered structure at Northampton, lifting the entire 1000-foot mass of wood bodily from its piers and abutments and dropping it into the Connecticut. There were six teams and eleven persons on the bridge at the moment it blew off, and it was a miracle that there was only one fatality.

Ithiel Town erected the well-remembered Springfield Toll Bridge in 1820, after a lottery had raised $25,000 to pay for it. Generations of Springfield people paid toll at the square portals to cross over the great arched spans. About 1878 a sidewalk was added, and there was talk of replacing the bridge. The authorities claimed then that it would "stand about ten years more." The bridge did far better than that: it passed the century mark and was not removed until 1922.

The old Springfield Toll Bridge spanned the Connecticut River for more than 100 years.

ISAAC DAMON

AMASA STONE, JR. and AZURIAH BOODY

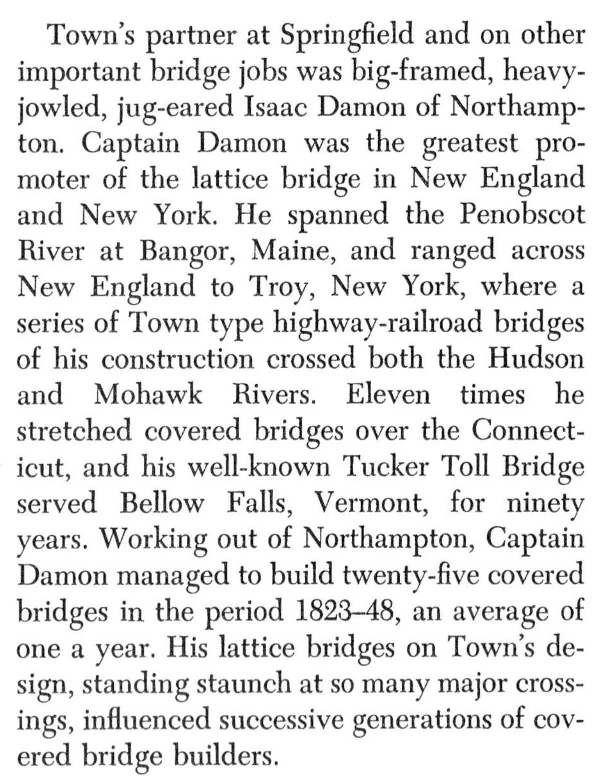

Town's partner at Springfield and on other important bridge jobs was big-framed, heavy-jowled, jug-eared Isaac Damon of Northampton. Captain Damon was the greatest promoter of the lattice bridge in New England and New York. He spanned the Penobscot River at Bangor, Maine, and ranged across New England to Troy, New York, where a series of Town type highway-railroad bridges of his construction crossed both the Hudson and Mohawk Rivers. Eleven times he stretched covered bridges over the Connecticut, and his well-known Tucker Toll Bridge served Bellow Falls, Vermont, for ninety years. Working out of Northampton, Captain Damon managed to build twenty-five covered bridges in the period 1823–48, an average of one a year. His lattice bridges on Town's design, standing staunch at so many major crossings, influenced successive generations of covered bridge builders.

Longest of any bridge over the Connecticut River was the original Western (later Boston & Albany) Railroad structure at Springfield. This bridge, 1330 feet long, was opened with a mammoth Fourth of July celebration in 1841, and was the first to carry iron rails across the Connecticut. It was a special design on the Howe truss plan and its construction was personally supervised by William Howe. Assisting the inventor were his partners and engineering friends who were soon to make Springfield the center of a network of bridge building companies.

William Howe had the good fortune to marry into a family who realized the potential value of his patent bridge and proceeded to make money for him hand over fist. The family was named Stone, and had four brothers who made the Howe truss bridge big business. They formed bridge building companies with overlapping ownership to contract for the truss in different localities. Amasa Stone, Jr., a cabinetmaker and church builder, teamed with an ex-schoolteacher and railroad brakeman named Azuriah Boody. Together they bought the New England rights to the bridge patent outright from William Howe for $40,000, but continued to pay him a royalty on every bridge erected. It was the start of a regular dynasty of bridge building firms, all made up of Howe and Stone relatives and their friends.

Brother Joseph Stone built in Maine, brother Daniel Stone had Pennsylvania and New Jersey, and brother Andros obtained rights to build the new bridge on the midwestern railroads. A friend of Boody's, Daniel Harris, came in with the original company to work Massachusetts and Connecticut, and Albert Briggs, later mayor of Springfield, took care of New York State.

RICHARD F. HAWKINS

Many of the Howe builders started young in the business. George B. Boomer from Sutton, Massachusets, was taken on as a junior partner by his older brother Lucius in the firm of Stone & Boomer. Railroads were demanding bridges so fast that young Boomer was sent west to open an office in St. Louis, Missouri. There, when he was only nineteen and not very big for his age, he was in full charge of thousands of dollars worth of bridge contracts and all their details. Although he took his work in stride, Boomer privately had this to say of himself: "It is shocking for a man to be so young—and short, too!"

Back in Springfield, sixteen-year-old Richard F. Hawkins went to work in the office of Stone & Harris in 1853. Eventually he succeeded to all the Howe bridge building companies in New England and put up highway and railroad bridges, both wood and iron, for fifty years. As late as 1885 he was still having his men erect an occasional tried-and-proved wooden Howe truss on branch rail lines in Vermont and New Hampshire.

Mr. Hawkins is remembered as a big man with a black mustache, who always drove a shiny black mare in his rounds of overseeing bridge jobs. He built an addition on his office expressly for the purpose of providing a hangout for old-time bridge builders. Springfield, being a center of this activity, had dozens of such men. They would "go down to Hawkins' " and their wives wouldn't see them all day. In the office they would sit and smoke and talk about how much better things were when they were young, and recount tales of building their favorite bridges. When he sold out to the American Bridge Company in 1900 Richard Hawkins was almost the last of the old cronies in Springfield who could remember when the Howe truss bridge was new.

The Greenfield area had a nest of covered bridges with half a dozen over the Green River and a big one over the Deerfield River near its junction with the Connecticut. This span was dubbed Cheapside Bridge in de-

Gilbertville Bridge looked like this before Ware got busy on its half with result shown on p. 32

rision at those who lived beyond it. In its last years it bore the legend "Erected 1806" on its portals, but the date is open to question. The Burr arch truss with which Cheapside was built was not invented until 1806 and not standardized until at least a decade later. A man named Consider Scott and one named Sheldon built on the arch plan in the Greenfield region about 1825–30. It seems more likely that Old Cheapside actually dated from the later time.

Just south of Greenfield an entirely different type of covered bridge, a real rarity, stood from 1906 until 1950. Its latticed trusses angled on a decided skew, and it spanned the main line of the Boston & Maine Railroad at the East Deerfield yards. It was an example of the very few covered bridges in America built as overpasses for railroad tracks. Local people knew it as McCallen's Bridge but it was carried on B&M records by the prosaic designation of A148.

The Boston & Maine also maintained a covered bridge over the Deerfield River, built expressly so that residents of Hoosac Tunnel, Massachusetts, could reach the railroad's depot on the far bank of the Deerfield River. This span, like any other railroad structure, was kept in shape by track gangs and was furnished with full water barrels at each end and a ladder to use in fighting possible fires.

A railroad was the cause for building a covered bridge near Athol. Whenever he went into the town Edmund Gage's horse was frightened by trains that ran parallel to the highway. Gage petitioned for—and obtained—a new town road and bridge across Miller's River well removed from the railroad track and its snorting engines.

Greenfield's new Pumping Station bridge, built April–November, 1972.

The Fitchburg line had bad luck with its covered bridges in this area. The very first train to Athol in 1847 never arrived in town. It crashed through a bridge into Miller's River, killing six men. At another crossing of the stream east of Athol a train was derailed when it hit a handcar thoughtlessly left in a bridge. The engine plunged through the floor onto the rocks of the river below, dragging the cars after it. This time there were three fatalities and twenty-five people hurt.

An encounter less tragic occurred in a covered bridge on the Pittsfield & North Adams Railroad. Pulling around a short curve into a dark span south of Cheshire Harbour, a P&NA train ploughed through a large family of skunks. The engineer made the mistake of stopping. Passengers, curious as to what had caused the sudden halt, jumped off the train in the darkened bridge amid half a dozen injured and infuriated wood pussies. That covered bridge, the coach and all the passengers took a long time to recover!

Cooley Bridge over the Chicopee River at Ludlow, Massachusetts, was involved with another kind of animal. A traveling circus passed through one night and there was a great discussion as to what toll should be paid for a group of camels. A fast-talking pitchman finally convinced the collector that, since camels were unlisted on the toll board, they could not be charged for.

Well over one hundred roofed spans were distributed over Massachusetts. The southeastern counties of the state, including Cape Cod, just never did have any covered bridges except for a railroad crossing or two. The tidal rivers of eastern Massachusetts are shallow and slow-moving and trestle and piling bridges sufficed the needs of local traffic.

Former Fort River Bridge in Hadley, once used by ghosts and anglers.

At Gilbertville over the Ware River stands a two-town bridge, where first Hardwick and then Ware each repairs its own half of the bridge independently and never does it at the same time.

In the Hockanum section of Hadley was the old Fort River Bridge, burned by vandals in 1962. The adjacent Connecticut had washed away the road to the north, so that it served only local farmers, fishermen and sweethearts who didn't mind a dead-end road. A ghost had been reported in this bridge, and the locality was certainly lonesome enough to support one. Another story concerns a Hadley taxpayer who thought the bridge should be replaced. Many years ago he reported at town meeting that he had felt the Fort River Bridge teeter perceptibly as his team passed over it on a trip to Holyoke. A man in the second row remarked dryly that the teetering probably occurred on the return trip from Holyoke!

Conway has a covered bridge across South River in the Burkeville section, and north of Greenfield the new Pumping Station Bridge crosses Green River in a beautiful woodsy setting. Off the Mohawk Trail is the last of the dozen covered bridges of Colrain, Arthur Smith Bridge over North River south of Lyonsville. This span was originally called Fox Bridge and stood south of Griswoldville. Abandoned when a new road was built, it was moved to its present site in 1896. Great laminated bows—added around 1920 so a cider company could transport heavier loads across the bridge—completely obscure the original timber arches. The span was put in good shape in 1956 by vote of the Town of Colrain.

Massachusetts is unique in having a number of comparatively new covered bridges. A long covered structure crossed the Deerfield River at Charlemont on the Mohawk Trail. The selectmen of Charlemont wanted to have this bridge replaced, and drafted an appeal in poetry to the Massachusetts Department of Public Works. It was entitled "Application for Old Bridge Assistance," and a portion of it went like this:

For a hundred years this bridge has stood,
(But for the past 15 has been no good).

Span over South River in Burkeville area of Conway.

And loving hands have patched its pants
'Til naught remains but food for ants.
By the Grace of God it stands today,
Exactly why no man can say.
But in the breeze its sways and jerks
Should give the chills to the Public Works.

At first it was planned to replace the Old Long Bridge in kind, but since it was wartime there were serious shortages in the nation's lumber supply. In 1944 a steel and concrete structure replaced the Deerfield River crossing in Charlemont.

The town still had a covered bridge, however: a small span called Bissell Bridge over Mill Brook up the hill from the village. Five years later it, too, was falling into ruin. The selectmen dusted off their old poem with new lines to fit the situation, concluding with:

The Bissell Bridge is falling down,
Right in the middle of our town
Please view the matter with alarm
And do vote "Yes" unto our plan.

The rhyme was a bit wobbly, but, according to the selectmen, not half so wobbly as the bridge, and the plan for which they sought approval meant a new covered bridge to replace the old one. State Public Works Commissioner William F. Callahan went along with their request. A muse in his department composed a proper answer:

Struck by the setting's natural beauty
The Commissioner said 'twas the state's duty
To save that lovely rustic view
And save the state some money, too.
For it seems the wood bridge can compete
And still be cheaper than concrete
The Town of course will pay its part
To gladden every tourist's heart.

The result was a fine new Bissell Bridge for Charlemont, built in 1951 by contractor T. J. Harvey of North Adams, Massachusetts. Charlemont held a square dance on the bridge in celebration of its opening, and it was such a gala affair that the custom is being continued on an Old Home Day each summer.

Old Sturbridge Village, one of Massachusetts' most famous tourist attractions, is a completely reconstructed community whose inhabitants perform the daily tasks typical of a century ago. It was natural for Sturbridge to want a covered bridge.

Earle W. Newton, then the director of the Village, dickered with the Vermont highway department to obtain a typical New Eng-

The new Bissell Bridge in Charlemont.

Vermont Bridge graces the exhibits at Old Sturbridge Village after its journey south from Dummerston.

land span. The one chosen was Taft Bridge on Vermont's Route 30 over Stickney Brook in the town of Dummerston. Wide enough for two cars to pass, Taft Bridge had done an unflinching job for eighty years. The bridge was presented by the highway department to the Village in 1951 and was dismantled piece by piece and trucked south to its new home in Massachusetts. George H. Watson, superintendent of reconstruction at Sturbridge, had a causeway built out into the Quinebaug River and placed the rebuilt Town lattice bridge on new abutments above a milldam in the stream. It served as a handsome exit from the Village for three years.

Then the hurricane of August, 1955, made the Quinebaug rise to unprecedented heights and floodwaters caused extensive damage at Sturbridge. Taft Bridge washed off its abutments and was saved only by ropes which snubbed it to trees on shore. A decision was made to move the bridge to a safer position where it could not be endangered. During the following winter the old span was moved a second time, to what is hoped will be its permanent site. Now called The Vermont Bridge, it spans an arm of the Quinebaug River between mill ponds.

In case you wonder how much money it took to accomplish the first-class job of relocating this bridge, the figures are available. Frank O. Spinney, Director of Sturbridge Village, reports that the total cost of bringing the bridge down from Vermont and re-erecting it at the first site came to $13,540.08. Replacing it on the new site after hurricane Diane cost another $12,000.

Another covered bridge forms the service entrance at Sturbridge. It was built under the direction of Superintendent Watson from old barn timbers, and, with its heavy beams and arches, one would never guess that this span is not every bit as old and authentic as any other relic among the exhibits. It was built in 1953 and its base is a former open bridge of slab concrete now completely hidden.

On a smaller scale was the newly built modern covered bridge southwest of Westfield,

Massachusetts. This was a well-made simple stringer affair erected in October, 1954, over Munn Brook in the town of Southwick. Fred and Seth Kellogg built themselves this 42-foot bridge on a lane leading to Seth Kellogg's home. They were tired of having open bridges rot away, they said, and they wanted a bridge that would last. But bad luck, in the form of the 1955 hurricane-flood, destroyed the bridge before it was barely ten months old. It was certainly one of the shortest-lived covered bridges on record, and photographs of it are already a rarity.

Two covered bridges representative of more than a century of building with wood cross the Housatonic River in Sheffield, near the western border of Massachusetts. At Upper Sheffield is an old narrow Town lattice bridge, dating from 1835. A little downstream is a startling sight: a big, new covered bridge, painted red, which was finished in 1953. This is the second of the "Massachusetts moderns," covered bridges designed for the highways of today but still retaining their old-time charm.

The new Sheffield bridge is 135 feet long and 24 feet wide, allowing three cars to pass abreast. It cost $250,000, and preparation of plans alone far exceeded the original cost of the older bridge upstream. It is a scientifically designed wooden panel truss devised by engineers of the Timber Engineering Company of Washington, D.C., and uses the company's patented metal connectors. The company calls this a Pratt truss, but it bears no resemblance to what T. Willis Pratt patented in 1844 and appears rather to be a sort of "multiple kingpost in reverse." The engineering company's trade name of *Teco* is an easier designation for this truss plan.

Like Bissell Bridge in Charlemont, Sheffield's modern span is the answer to a petition by the people of the neighborhood for a bridge to be replaced in kind. Thus the tradition of covered bridge building is being carried into the Jet Age, and it seems to promise that there may be some roofed spans still around to edify and delight several future generations.

Interior and skeleton views of the Bissell and Sheffield Bridges, two "Massachusetts moderns," display Jet Age wooden trusses devised by the Timber Engineering Company and using Teco *connectors.*

CONNECTICUT:
Birthplace of Builders

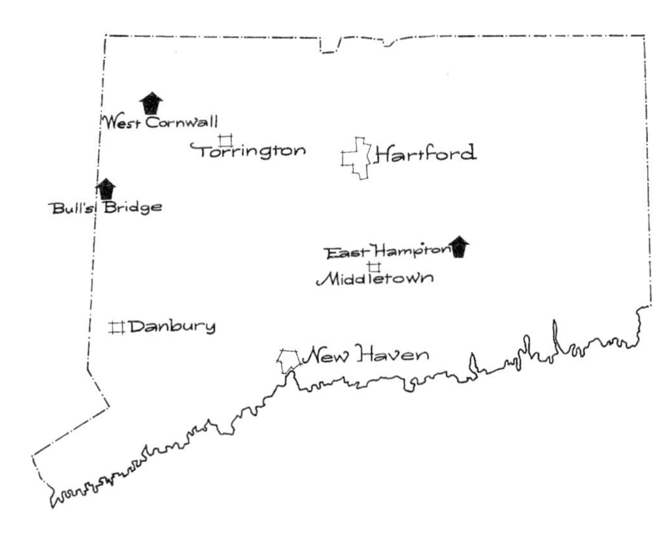

MEMBERS OF THE older generation, as well as other people with an eye to the picturesque, have for years mourned the gradual disappearance of the covered bridge in Connecticut. There are still three tunneled wooden spans in the state; two in everyday use and one honorably retired.

Although all three have stood for many years, some Connecticut folks are unaware of their existence and believe that it is necessary to journey to Vermont to see a covered bridge. Even less known is the fact that over sixty covered highway and railroad bridges once stood in the Nutmeg State. These were scattered for the most part down the valleys of western Connecticut, crossing the Housatonic, the Naugatuck and the Farmington Rivers. The other half of the state appears to have had very few, with only a dozen known to have stood east of the Connecticut River.

Connecticut's chief contributions to the annals of covered bridge building were men. The well-known early bridge designer, Theodore Burr, came from Torringford, a hill town that flourished when present-day Torrington was a swamp. Architect Ithiel Town, of lattice bridge fame, hailed from Thompson and made his home in New Haven. Then there were master builders like Jonathan Walcott of Windham, Colonel Ezra Brainerd of East Hartford, Zenas Whiting of Norwich and Samuel Mack of Lyme. Oddly enough, most of these men gained their greatest fame outside Connecticut. Walcott built the longest covered bridge in the world, one 5690 feet long across the Susquehanna River in Pennsylvania from Columbia to Wrightsville, and Burr crossed the same river at McCall's Ferry

This picture immortalizes the passage in 1892 of the first trolley car over historic Hartford Bridge.

with one 360-foot span, the longest single-span wooden bridge ever erected.

Walcott's first bridges were uncovered arched affairs. One across the Connecticut River at Hartford was erected and operated by a private company, with the toll house high and lonesome out in the middle of the span. This pioneer structure succumbed to ice and high water after eight years and the first known covered bridge to be built in Connecticut took its place. Long a landmark at the capital, the Hartford Toll Bridge was built in 1818 by Ithiel Town of New Haven and Isaac Damon of Northampton, Massachusetts. This massive bridge was 974 feet long, built of native pine and divided in the center so that the traffic using it was separated. The dark bulk of the bridge was partially illuminated by a long cupola on the roof of the two center spans, with skylights in its peak. At night oil lamps were strung at intervals across it.

One of the first custodians of these lamps was a boy named Thomas Martin, who kept them filled, trimmed and lighted seven nights a week. He rose to become president of the Hartford Bridge Company which owned and maintained the venerable structure. Tolls were abolished in 1889 after seventy years' return on the original $40,000 invested by the stockholders. The bridge was purchased and made free by the five towns which stood to benefit most from the abolition of tolls. A few years later the dark tunnel was laid with rails, and on March 28, 1892, the first little trolley car ventured through.

One of the early bridges on Town's "mode" crossed Farmington River at Windsor.

A tale which has a ring of truth concerns the sign which once hung on the portal of the Hartford Bridge. This sign differed slightly from the usual "walk your horses" admonition. The story goes that a farmer's wife from over beyond East Hartford was in a hurry to get to market on a Saturday morning, and drove a bit too fast to suit the conscientious constable stationed on the old bridge. Hauled into court, the lady was about to plead guilty and pay her fine when a smart lawyer interceded, leaping to the astonished lady's side.

"May it please Your Honor," he shouted, "I move this case be thrown out of court!"

"How so?" asked the judge.

"Your Honor, the sign on the Hartford Bridge reads 'Ten dollars fine for any man to ride or drive faster than a walk on this bridge.' Your Honor, the horse was a mare and my client is a woman!"

In the uproar that followed the case was dismissed.

Fire completely destroyed the Hartford Bridge on the night of May 17, 1895. The flames started near the East Hartford end and in less than ten minutes were sweeping through the entire tinder-dry structure. There had been much agitation for a new bridge, and among Hartfordians there were few to lament its spectacular passing. The *Hartford Courant* reported that a crowd of 20,000 persons lined the banks—everybody not bedridden or in jail—to see the bridge burn. The great wooden arches, it was reported, "glowed like brilliant blazing rainbows before crumbling slowly into blackened embers that drifted gently seaward on the crimson-glowed surface of the river they had spanned so long."

Squared portals on Farmington's bridge copied Town's drawings but they were unusual on a Yankee span

A hose cart and two of Hartford's pet fire horses from Engine Company No. 3 were trapped inside on the south lane of the bridge, and went into the river when it fell. Some enterprising citizens recovered the carcasses and sold the twisted, rusty horseshoes as souvenirs. The first eight were quickly purchased; the demand increased, and it was said that neighborhood barns sported enough tacked-up "shoes from the brave fire horses that perished in the great Hartford Bridge fire" to have shod a cavalry regiment.

The first of the well-known lattice covered bridges in New England was built in 1823 under the direction of Ithiel Town, inventor of the truss. He supervised the erection of one for the New Haven and Hartford Turnpike Company over Mill River at Whitneyville. The bridge stood on this site until 1860, when Eli Whitney II created a bigger reservoir in the river. Although engineers held that the structure couldn't be moved, Whitney jacked up the turnpike bridge and hauled it on rollers half a mile upstream to a new site at Hamden. There it stood for another thirty years.

The Enfield-Suffield Toll Bridge across the Connecticut River had a life nearly as long as the Hartford Bridge enjoyed farther downstream. It spanned both the river and the adjacent Enfield Canal with a thousand feet of lattice truss. A flight of steps led up from the towpath to a unique side entrance for foot passengers using the bridge. It was a lucrative piece of property for the Dixon family, who owned it for many years. The New Haven and Hartford Railroad Company had to pay $10,000 for the privilege of crossing between towns where the Dixons held exclusive bridge rights. Old age and lack of maintenance finally did in the Enfield-Suffield Bridge on February 15, 1900, when a freshet tore away three spans and bore them off down the Connecticut. On the bridge at the time was Hosea Keach, the railroad agent at Enfield. Word was telegraphed down to the iron railroad bridge below; there, after a nerve-wracking three-mile ride on the floating wreckage, Hosea was thrown a rope and

Unionville's bridge over the Farmington was always decorated on Memorial Day and Fourth of July.

Bull's Bridge over the Housatonic River.

pulled to safety.

All covered railroad bridges in Connecticut have long been replaced. Most of them served the predecessors of the New Haven Railroad. Big high-sided affairs with heavy timbers and thousands of feet of siding, they stood at such places as Greenwich, Bridgeport, Milford, Waterbury, Norwich and Jewett City. At each was stationed a watchman, usually an elderly or disabled railroad employee, to inspect the bridge for live sparks after each old-time wood-burner had chuffed through. Collinsville's railroad bridge, on a branch line, was a great place for pigeons. The birds were seemingly unmindful of the smoke from an occasional train. Local boys climbed up into the eaves and made a business of carrying home dozens of squabs each season.

On the railroad line along the shore of Long Island Sound the covered bridges had to be renewed early—not as the result of structural failure but because of ravages by teredos. These little marine worms ate away the untreated wooden piling and cribwork from under the bridges. At Saugatuck a covered span suddenly dropped a foot as a train rumbled over. Fortunately the rails held and there was no accident, but the bridge was sprung so badly that it had to be replaced. Presoaking the piling in tar solutions eventually overcame the extensive teredo menace, but extra caretakers in small boats were continually probing the wooden underpinning for signs of danger. This was a full-time job on a bridge like the large crossing on the main line of the New Haven between Devon and Naugatuck Junction. This, the longest covered bridge to be built in Connecticut, stretched 1080 feet across the wide Housatonic River near its mouth, with seven covered spans and a draw so that ships could proceed further upstream.

Only two of the eighteen covered bridges which once spanned the Housatonic River within the State of Connecticut remain today. It is fine that the state has seen fit to help the towns repair and maintain these bridges. Now succeeding generations may not only enjoy the sight of them, but acquire the respect and admiration for sound workmanship in wood which comes with actual day-to-day use of a covered span.

West Cornwall Bridge was built well over one hundred years ago, and was once called Hart's Bridge after an early settler who lived near this crossing of the Housatonic. Tradition has it that Selectman Marcus Smith journeyed to North Adams, Massachusetts, to choose the timber for the bridge personally. After a century of rough usage, including the buffeting from a dozen big floods, the bridge was in an advanced state of decay. Then, in 1945–46, the state stepped in and gave the

Comstock's Bridge is ideal for picnics or a wedding, but West Cornwall Bridge continues its job of carrying traffic across the Housatonic.

old bridge a thorough overhauling. A typical village bridge, the West Cornwall covered span leads across the river to the new "Covered Bridge Shopping Area" which has been built in West Cornwall. Occasionally, of a summer evening, the bridge is roped off for a neighborhood square dance.

Between Kent and Gaylordsville the Housatonic is spanned by Bull's Bridge. This crossing served a new route to the Hudson River over which the products of Jacob Bull's iron furnace were hauled to New York markets. It was built in 1842 at a cost of $3000. When the Connecticut Light & Power Company built a power plant at the site, Bull's Bridge was jacked up twenty feet, and has stayed well out of the reach of flood waters ever since. A renovation job in 1948–49 left it trim with new internal bracing and siding and a coat of colonial red paint.

Far to the east stands Comstock's Bridge, spanning the Salmon River just off Route 16 between East Hampton and Colchester. Twenty years ago it was by-passed by a new concrete bridge. The Civilian Conservation Corps boys rebuilt the original covered span, fitting it with ornate doors and glass windows. A piece of a replaced main stringer is on exhibit in the dark interior, demonstrating how the unknown builder compactly lapped and bolted his timbers. Also notable is Comstock's thirty-foot approach span. This wooden bridge, with the low trusses enclosed on both sides and the floor exposed to the weather, gives an idea of the many open bridges of the unroofed variety which once spanned Connecticut streams. Since it is no longer used for highway traffic, Comstock's Bridge has become in late years a favorite place for local picnics and lodge outings. Even a wedding once took place within its sheltering portals.

Remains of the Enfield-Suffield Toll Bridge in 1900 shortly after Hosea Keach's ride.

RHODE ISLAND: Only Memories

UNLIKE ITS NEIGHBORS in New England, The State of Rhode Island and Providence Plantations—as it is named officially—never made use of covered bridges to any great extent. Rhode Island's wooden bridges were for the most part either pile structures over the tidal streams that border Narragansett Bay or simple uncovered truss spans over the inland brooks. Large covered trusswork bridges were unnecessary because the rivers were neither swift nor deep.

Although there were doubtless others, careful research has disclosed records of only five covered bridges which stood in the state—two highway spans and three railroad bridges. They stood out because they were uncommon.

First was the Washington Bridge across the Seekonk River from Providence to the Massachusetts shore. The original open bridge here, to the east of the city, was built in 1793 to replace a ferry. John Brown, active in early Providence affairs, organized the Providence South Bridge Company to build it. Brown was a friend and admirer of George Washington, and on his new bridge he placed a marble tablet inscribed "as a testimony of high respect" for the first President. A specially-commissioned wooden statue of the Father of His Country also adorned one end of the first Washington Bridge.

A storm swept away the first crossing at this spot and Providence's great gale of 1815 did the same to the second one. The statue of Washington went with the bridge and was never recovered, but the marble tablet still exists as an exhibit of the Rhode Island Historical Society. Ferry service was re-established for a time and then, some time after 1820, the substantial covered Washington Bridge was built. This was described as a "long wooden bridge and three quarters of its length covered by a house." It appears to have been a double-barreled lattice truss bridge on the plan of Ithiel Town, who may well have had a hand in its construction. It had four spans and a little open draw on the eastern end.

As a toll-collecting venture the Washington Bridge proved a good investment to its stockholders for over eighty years. Tolls were finally abolished, and it became state property; it was replaced by an "up to date" structure in 1884. During its lifetime the old crossing had ceased to be an interstate bridge, for in 1862 the state boundary was moved eastward, putting East Providence in Rhode Island.

Right alongside the Washington Bridge, and doing an equally good job, was the India Point Bridge built by Thomas Hassard strictly for railroad use. For a time the Boston & Providence Railroad carried its passengers in cars drawn by horses from the old depot in Provi-

A predecessor of the old Washington Bridge over the Seekonk River between Providence and East Providence sported a statue of the first President.

dence across the Washington Bridge to East Providence, where the engines were coupled on. This state of affairs was ended by building India Point Bridge across the Seekonk in 1835. It was the first interstate railroad bridge in the United States.

Hassard was a handsome Irish carpenter who specialized in bridges for the infant railroads, using the patent plan of his old friend and mentor, Colonel Stephen H. Long. The Seekonk Bridge was one of twenty-three he built for six different pioneer rail lines over a period of six years. Hassard's structure at Providence survived heavy traffic and was busy even when the line across it became only a branch. After the Civil War a new bridge with a 60-foot draw was built to replace the original. The new one had a Howe truss put up by the patent-right holders for Southern New England—Daniel Harris and Richard Hawkins of Springfield, Massachusetts. This second covered bridge carried the freight trains over to India Point until 1901.

The only other covered highway bridge known in Rhode Island—aside from Washington Bridge—was down in Washington County near Narragansett Pier. It spanned the Pettaquamscutt (or Narrow) River on the Boston Neck Road (now U.S. 1) north of 'Gansett. This, too, was a Howe truss, built across the tidal stream in 1866–67. It had unusual portals cut neatly in the form of deep half circles.

This isolated covered bridge stood for half a century, and about the only time it made headlines was near the end of its existence. The old South County landmark had become a bit unsightly through the years and it was claimed that "the only transportation facility that carried more advertising was a London bus." Lovers of rural scenery declared that the highways were getting besmirched by advertising posters, and pointed to the Narrow River Bridge as a horrible example.

These individuals found a spokesman in Charles C. Blanchard of Providence. Armed with an axe, he drove his runabout down to the bridge one sunny October afternoon in 1915 and began hacking off the offending signs. Some were nearly as old as the bridge itself and advertised everything from Prince Albert Tobacco to the Great Kingston Fair. Assisting Mr. Blanchard was Edward S. Cornell, secretary of the Highways Protective Association, carrying a clawhammer and a crowbar. The zealots had before-and-after photos taken, and there was certainly a great improvement in the looks of this segment of the Boston Neck Road. Blanchard's and Cornell's blows for beauty were the beginning of the end for Rhode Island's last covered highway bridge. State highway crews demolished the old span in favor of a wide concrete structure in 1920.

After the Narrow River Bridge was replaced it seemed that Rhode Island no longer had any covered bridges. Yet up in the industrial heart of busy Woonsocket was one that had been overlooked. This was the roofed and weatherboarded structure on the Hamlet Branch of the New York, New Haven and Hartford Railroad, crossing the Blackstone River to the old Social Mills.

The Hamlet railroad bridge was unique in many respects. It was put up by the New Haven road in 1898 as a regular open wooden

Two views show Rhode Island's Hamlet Bridge before it was destroyed by Hurricane Diane in 1955.

truss, and its covering was added at some later date. Since the bridge crossed the river at an angle the truss was on a skew, giving the roof an odd lopsided appearance, like a dowager with her bonnet awry. Attached to the downstream side was an open walkway, used as a short cut by people going to and from work in the mills and by ever-present boys.

Once a day a New Haven diesel switcher rumbled over the 127-foot Howe truss span with freight to pick up or deliver at the American Wiper-Waste mill on the north bank of the Blackstone. A *Providence Journal* feature story in 1951 brought the little railroad bridge into the limelight. It was listed as a point of interest in state publicity releases and spotted on official state highway maps, standing as a strange example of rural charm only a stone's throw from the bustling streets of Woonsocket.

The Hamlet Bridge was entirely adequate for its once-a-day use by the light switcher and a few cars, and the New Haven expected to go on using the bridge for years. But Fate had other plans. The terrific rains brought on by Hurricane Diane in August, 1955, raised the Blackstone River to heights never seen before. The rampaging river tore away the Hamlet span and left Rhode Island the only New England State without a covered bridge.

NEW YORK: Gateway to the West

NEW YORK STATE was once a vast proving ground for covered bridges. Big and little bridges of every known type were thrown across streams from Sing Sing to Massena and from Granville to Gowanda. There are records of more than 250 of the bridges, with most of the best-remembered ones in the Hudson, Mohawk, Delaware and Susquehanna Valleys, and a goodly portion in the Great Lakes-St. Lawrence River area.

Many people think that because pioneers from New England settled much of New York they brought the idea of covered bridges with them. In reality, most of the early settlers had never worried much about bridges until after they went "west" to York State. But this new country was different. There were so many rivers that they had to build bridges if any kind of communication was to be established; and they set to work.

Theodore Burr of Oxford, who commenced spanning streams in 1800, was the dominant figure of early bridge building in New York. His experimental structures over the Mohawk River at Utica, Little Falls, Canajoharie and Fort Plain were widely seen. Nearly one hundred and sixty-five years ago he spanned the Mohawk at Schenectady with an unprecedented *suspension* bridge of wood—a unique conglomeration of mammoth supporting beams and long laminated cables of built-up

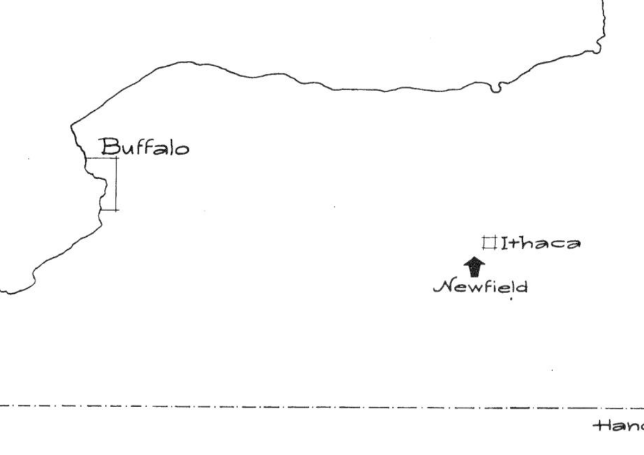

flat plank from which hung the floor timbers. This contraption stretched like an undulating snake across the river; and, although they are built on variations of the suspension principle, even the graceful hanging webs of the George Washington and Golden Gate Bridges owe something to Theodore Burr.

It was his more conventional arch-truss bridges that caught on, however. Any good house- or millwright who crossed the Hudson by way of Burr's big span at Waterford was sure to stop and take note of its construction. The turnpike companies, building roads across the wilderness, needed bridges at all their major river crossings. Theodore Burr and his followers obliged them. They were Jacob Coss of Bainbridge—described as "a man so rugged he chewed tobacco and swallowed the juice"—Reuben Field, who worked in the Hudson Valley, and Daniel Stewart of Saratoga Springs who spanned the Sacandaga River. Down south of the Catskills, Major Salmon Wheat had contracts for the bridges on the Newburgh and Cochecton Turnpike which was being pushed through, nearly string-straight, across the rough Shawangunk Mountain ranges.

Major Wheat's masterpiece was a single-arch bridge over the Neversink River at what became known as Bridgeville. This 160-foot double-barreled wonder was built in 1807, and was the first known covered bridge in New York State. Wheat was ahead of Burr in closely weatherboarding his trusses, and he gave the bridge ornate portals with wide wing-walls and the unusual feature of a sidewalk in the *center* between the roadways. It is a tribute to Salmon Wheat that Bridgeville Bridge lasted 116 years; even today the stubs of the mighty arches can still be seen, protruding from the old abutments where they were cut off in 1923.

The first New York State wooden bridge on which Theodore Burr himself put a roof was long-lived too, beating out Bridgeville by three years. This was Esperance Bridge over Schoharie Creek, west of Albany on the Cherry Valley Turnpike. It is said that Burr started this bridge in 1809 and then went home to Oxford to build a fine house for his growing family. The bridge company had just about given him up when he returned jauntily two years later, and finished the bridge. With its three massive arches, the old

Theodore Burr's wooden suspension bridge built in 1809 over the Mohawk at Schenectady was not a success, yet its sagging trusses—later piered and housed with hodgepodge roofing—held until 1872.

Esperance crossing was a Schoharie County landmark from 1811 to 1930. In the present century, when automobiles began to use the turnpike, drivers were often startled when the bridge loomed up before them at the foot of a long dark hill from the east. One man, thinking the bridge was a big barn, drove right on straight—into the creek. Horizontal siding, like that on Esperance, was used on sizable spans at main cities and turnpike crossings and became the trademark of an "old" bridge.

A great number of the early York State spans were of the double-barreled variety. One of these was proposed for crossing the Hoosic River northeast of Troy. The Hoosick & White Creek Bridge Company (spelling varied from 'way back: the river is Hoosic, the town is Hoosick, the tunnel is Hoosac) was incorporated to carry out the project, and in 1827 it built a fine big two-lane covered bridge. A neighborhood hunter shot an eagle one day, and to show off his prowess he nailed the carcass over one entrance to the bridge. Eventually the bird rotted away, and a young girl who lived nearby climbed up and painted an eagle in its place. The colorful design was admired, and soon the town that had sprung up around the crossing began to call itself Eagle Bridge. Other bridges have been built on the site and each has carried on the tradition. Grandma Moses, the primitive painter, has put her recollection of the structure and its decoration on canvas. The modern bridge of today still sports cast-iron eagles taken from the second bridge, high up on its steel crossbeams.

Isaac Damon from over in Massachusetts and Joseph Hayward of West Troy (Watervliet) built the first railroad bridge across the Hudson River in 1835. One lane carried the tracks of the Rennselaer & Saratoga Railroad out of Troy, on the other was a highway. Lacking a roundhouse, the little railroad company often stored its engines in the big bridge. This was a fatal error. On May 12, 1862, the bridge caught fire from the sparks of a locomotive and the resulting conflagration took most of downtown Troy. Five hundred and seven buildings were destroyed at a loss of three million dollars—the worst fire in the city's history.

One double-barreled wooden bridge won a contest with an iron one. At Fort Plain the local folks were unhappy about paying high tolls to cross a long covered span across the

Mohawk. After agitating unsuccessfully for a free bridge, a group got together and had an iron one built right alongside the toll bridge. There were injunctions and lawsuits galore, and at last the toll bridge company gave up and went out of existence. The two structures carried on side-by-side for several years. Then the iron one was sold to raise money for putting the wooden one in good shape!

The country turnpike bridges didn't see as much travel as their city and village cousins did, particularly after the railroads drained their traffic away. On the Western Turnpike out of Albany, the highway crossed the Norman's Kill on a short double-barreled bridge whose middle partition was a bit off-center. It is rumored that a neighboring farmer fenced off the narrower portion of the little used bridge to serve as a pig pen. The odor was probably enough to make the occasional traveler hurry through it much faster than a walk.

In the western part of the state Alexander McArthur and John Mahan went after bridge contracts with a will, building railroad spans, highway bridges and some Erie Canal farm crossings in the Genesee country. They were aided and abetted by Dr. Moses Long, who had moved from New Hampshire to Rochester to carry on the promotion of his brother's patent truss. Best-known of the Long bridges built by this combination of talents were the span at Squakie Hill (Mount Morris) over the Genesee River and the covered bridge in Rochester which carried the little Tonnewanta Railroad and its "elegant passenger cars" across the Erie Canal.

Although his work was mostly with iron bridges, an adopted York State engineer-inventor should be mentioned here. Squire Whipple has been called "the retiring and modest mathematical instrument maker, who, without precedent or example, evolved the scientific basis of bridge building in America." Squire (this was his given name) moved with his family as a boy from his birthplace in Hardwick, Massachusetts, to the Otsego County town of Springfield, New York. He spent four terms at the Hardwick and Fairfield academies, taught school, farmed and studied all manner of subjects at home. In 1829 he was able to enter the senior class at Union College and graduate the next year. He did engineering work on the Baltimore & Ohio Railroad and the New York State Canals, became interested in bridge design and patented an iron bowstring truss for bridges. Designing for wooden spans and later for iron bridges, Squire Whipple also invented what he called his "trapezoidal truss" in which the heaviest bracing inclined toward the ends, rather than toward the center. He said that bridge planning of his day had so far consisted only of "cut and try." If a bridge stood there were no questions asked. To his mind, rational methods of design for bridges had to be developed if the extension of railway lines

First known covered bridge in New York State (1807), Major Wheat's span at Bridgeville stood until 1923.

SQUIRE WHIPPLE

was to continue unhampered. Squire Whipple's big contribution was a book published at Utica, New York, in 1847, and titled *A Work on Bridge Building . . . and Practical Details for Iron and Wooden Bridges.*

It took a good many years for the importance of Whipple's writings to be appreciated. Through them he became recognized as the first man ever to analyze correctly and adequately the stresses in a bridge truss—i.e., the various "pulls and squeezes" by which the weight of a load is carried from beam to beam and finally delivered to the solid abutments. After Whipple bridge building became a profession, rather than a trade. He acted as consulting engineer for a few covered wooden spans built in New York State—two over the Mohawk River at Schenectady and

This is one of Whipple's truss designs for wood, used in some York State covered bridges.

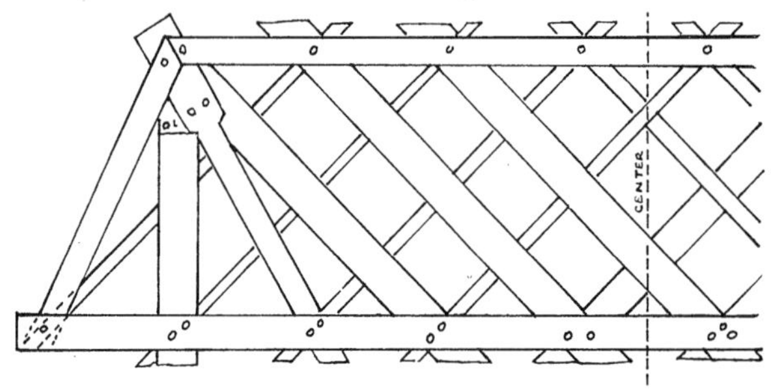

Cohoes, and one over West Canada Creek below Trenton Falls which stood until 1937. Whipple planned the little open queenpost-truss bridges over the Erie Canal, competing against a design by Colonel Long to win their adoption. They were built at hundreds of farm and country road crossings along the length of the waterway.

This thin, gray-bearded bridge inventor and writer lived a quiet and simple life in Albany, manufacturing and selling surveyor's instruments. After many years his genius for calculating the stresses of bridges was recognized. Honors came to the old civil engineer from colleges and professional societies, and his famous little book was in print for fifty-two years. But Squire Whipple was content to spend his last days being, as he put it, "an insular atom of human nature, only connected with the mass by a few . . . bridges."

The greatest number of covered bridges in New York State today is concentrated in what some call the "backside of the Catskills" in Delaware and Sullivan Counties. There are still three over the East and West Branches of the Delaware River, a sizable stream even here where it is just starting its long journey. Well-known to fishermen and campers are the bridge at Beaver Kill Campsite and the two covered spans across the trout-filled Willowemoc.

Delaware County's most famous bridge builder was a precise and able Scot named Robert Murray. Murray had come to this country as a boy; he tried storekeeping and then took up heavy contracting. Between 1854 and 1859 he built four Long truss bridges over the East and West Branches of the Delaware River. As he worked down the West

The covered bridge in this old picture was built by Obadiah Wilcox over the Sacandaga at Hadley in 1813. Beyond, the Adirondack Railroad's first engine, Maj. Gen. Hancock, *perches on an open deck bridge.*

Branch, he was careful to build each bridge ten feet longer than the one above to allow for the widening stream. This precaution has helped save two of his bridges—at Downsville and Hamden—down to the present day.

Later years found Robert Murray building Town lattice bridges, and he even tried an experimental Haupt truss or two which he'd read about in books on carpentry. It is related that Murray, who lived in the hill town of Andes, would walk to his bridge sites early Monday morning, board out during the week, and hike home again Saturday night. During these weekly trips he always went barefoot, carefully carrying his shoes. Thinking about Mr. Murray's Scottish background, there may be some truth in the story!

This region is also noted for its "migrating covered bridges," ones which began life in one spot but are now to be found in another. Back in 1870 the town of Delhi had to replace several covered bridges taken out by a disastrous flood. Among the new spans was a bridge built by James Frazier and James Warren on Kingston Street in the village. After fifteen years, around 1885, a company sold the town fathers on putting up a "modern" iron bridge to take its place. It seemed a shame to destroy a wooden bridge so comparatively new. Instead, David Wright and a town crew set to work, carefully marking the timbers and taking the bridge apart. Next it was hauled in wagons to East Delhi, three miles up the West Branch. Re-erected, it is still called Fitch's Bridge, after an old name for the crossing. Above the red clay dust which stains the floor can be seen the letter-and-number markings on the old lattice timbers. Mr. Wright didn't want any mistake made about which were the top chord pieces: they are painted with the word *Upp*. He also found it handier to re-erect the bridge to face in the opposite direction from which it had stood in Delhi. As a result, all the timbers marked *East* are on the west

Robert Murray and his bridge at Hamden using Long's panel truss.

Fishermen and campers still enjoy the Beaver Kill Campsite Bridge, built by John Davidson in 1865

side, and vice versa.

Joseph Sherwood of Livingston Manor salvaged a covered bridge in a different manner. When the state highway was put through town, he dismantled the structure at the Manor and moved just half its length to a new site upstream. That was all he needed to cross the narrowing Willowemoc Creek beyond Debruce. There the span stands today and is known as Bendo Bridge.

Over on Mill Brook in Delaware County a private organization, the Tuscarora Club, owns the stream and its fishing rights. When a covered bridge at Dunraven was to be replaced by a new structure in 1935 the club dismantled and rebuilt the span over a beautiful waterfall and pool in Mill Brook. It is used as a footbridge by club members and is even fitted with steps on one side. Still another small bridge over Lew Beach Hill Brook at Lower Shavertown was slated for early demolition as the Downsville Dam neared completion. Water for thirsty New York City would soon flood the whole valley of the East Branch, backing up nearly to Margaretville. Carl Campbell, a lumber dealer in Roscoe, bought the bridge and trucked it in sections over the mountains to Methol, where he owned property. Cut down a bit and re-erected over Trout Brook, the old Lower Shavertown Bridge makes a fine entrance to Mr. Campbell's hunting lodge.

Former bridge over the Hudson at Glens Falls.

Migrating in 1934, a bridge from Turnwood was moved without fanfare across Ulster County and re-erected on a private estate over the old channel of Esopus Creek below the Ashokan Dam. In fact, so quietly was this done that it took covered bridge enthusiasts over twenty years to get wind of the bridge and "discover" it!

Perrine's Bridge near Rifton is Ulster County's best-known bridge, and it is probably seen more often than any other covered bridge in the state. It spans the Wall Kill adjacent to the New York State Thruway. It is surprising how many people spot this covered bridge and remember it, even though it can be glimpsed for only a brief moment by travelers in the northbound lanes. Built in 1844, Perrine's Bridge is the only sizable example of the Burr arch truss still standing in this state where it was invented.

Oldest in New York is a small sturdy arch bridge in Glimmerglass State Park, at the north end of beautiful Otsego Lake. This bridge, now state property, spans Springfield Creek (also called Shadow Brook). It was erected between 1815 and 1830 by George Clarke, who established a country estate here during that period, called "Hyde Hall." Among old private bridges is a little one at Edinburg, in Saratoga County. Built for, and used mainly by the cows of farmer Arad Copeland, this bridge dates from 1879 and spans a former mill pond in Beecher Creek.

Back in 1884 the Oakwood Cemetery Association of Troy wished to provide a new entrance to its grounds. The cemetery occupied a sidehill bordered by the Fitchburg Railroad tracks and the only safe crossing meant a new bridge. Extending from 101st Street, a new earth ramp and a covered Howe truss highway bridge were built to clear the railway line. This bridge, built on a skew, had ornate portals and was painted dark green to harmonize with its surroundings. For protection against sparks from the old steam engines, it was sheathed top to bottom with corrugated iron. Unhappily, this did not save the structure from a deliberate fire set by unapprehended vandals. It was the last covered wooden railroad overpass in the United States. the hills and valleys resemble those of neighboring Vermont. Buskirk's Bridge, the last remaining covered span across the Hoosic

Four typical New York spans are: Fitch's Bridge, a migrant come to rest at East Delhi; the Lower Shaverton Bridge which was dismantled and trucked over to Methol; the fine private crossing recently built in Albany County by Gerald Waldbillig and his son, and the state's oldest covered bridge (c. 1830) which is now in Glimmerglass State Park north of Otsego Lake.

America's last covered railway overpass, the Cemetery Bridge spanning B&M rails in Troy, was victim of casual vandalism.

River, is at the county line, and there are three sizable bridges across the Batten Kill to the north of Cambridge. Since the Kill is a famous trout stream, the bridges at Eagleville, Shushan and Rexleigh are well-known to anglers. It is said that when the Shushan Bridge was completed in 1858 its builders, the local Stevens boys, celebrated in style by leaping exuberantly off the roof into the millpond which backed up under the brand-new span. Today, Shushan Bridge has been preserved in an odd manner, perched high and inaccessible on a concrete pedestal.

At Newfield, near Ithaca, is the only covered bridge in New York State's western section. Details of the contracting for the Newfield Bridge are of interest. This 80-foot span was first planned in 1848 but was not completed until three years later. Lumber was purchased for $6 per 1000 feet and one man, Daniel Tunis, turned all the wooden pins with which to secure the Town lattice truss. Masons Ben Star and Dick Russell laid the abutments. Carpenters were Samuel Ham and his sons David and Sylvester, David Dassance and Patchen Parsons. The workmen toiled for $1 a day from sunup to sunset. The complete original cost of Newfield Bridge, now well over a century old, was only $600!

Arson also removed Cowlesville Bridge from its rural scene.

By far the most notable covered bridge in New York is the Old Blenheim Bridge over Schoharie Creek at North Blenheim, built by Vermont's Nicholas Powers on a unique variation of Long's plan. This is the longest single-span covered bridge in the world, and is built with a type of construction never attempted elsewhere. It is also one of only six double-barreled covered bridges still standing in America, and the only one in New York with a divided lane.

The Blenheim Bridge Company was formed early, but nothing was done about building until 1854. Then Major Hezekiah Dickerman, a Connecticut man who ran the local tannery, was elected supervisor of the town. Major Dickerman decided there was enough money in prospect to go ahead with the span over the Schoharie, and he got work on the stone abutments under way while he was still looking around for a good contractor. It is related that the major was a stern old New Englander and death on liquor. One morning his masons had some liquid refreshment on the job when Hezekiah drove up unexpectedly. The workmen hastily stoppered their jug and thrust it into a cranny of the half-finished abutment. Dickerman had figured to complete that section that very day, and he stayed to supervise personally the laying of the mortar. So it is presumed that in one of the abutments of Blenheim Bridge lies a jug, and its mellow contents

Flame does quickly what time alone could not accomplish on old Cameron Bridge near Gravesville. The structure was abandoned in 1937 and then purposely set afire to fall into West Canada Creek.

are vintage 1854.

The company engaged Powers to make the wooden superstructure of the bridge, and he turned out a wonderful piece of work. He devised a bridge of three trusses on a modified panel plan, with huge pine timber posts, braces and counterbraces bolted together. The center truss is higher than the two outside ones, and in this center truss he enclosed a single arch of oak which reaches from below the lower chord on up to the ridgepole at the peak of the roof. It is really three concentric arches, one on top of another, and carefully blocked apart to allow air to circulate among them. Powers had the bridge timbers joined tightly together with 3600 pounds of wrought-iron bolts and about 1500 pounds of washers.

Powers received $7 a day to superintend the building of this giant among covered bridges. He had his men frame the parts on the flats back of North Blenheim village and then put it together over scaffolding in the creek. The covered portion of the bridge is 232 feet long, the trusses 228 feet and the clear span 210 feet between abutments. The two lanes are 26 feet wide.

When the new steel bridge on the state highway was erected nearby in 1932 Old Blenheim Bridge, long since free from tolls, was taken over by Schoharie County and preserved for the public as an historic exhibit. Since then the bridge has been kept in hardy retirement. In 1954 $5000 was spent in refacing the eastern abutment (no, they didn't find the jug!), and 1955 saw a full-scale celebration in honor of the bridge's one-hundredth birthday. Visitors came from a dozen states to watch a mile-long parade and take part in the anniversary ceremonies. Flag-draped, the old bridge stood serene through all the band music and speech-making. Today it is a fine picnic site visited each year by hundreds of people who walk across the long span and pause in the dim interior to admire Nicholas Powers' skillful bracing and mighty arch.

Old Blenheim: staunch crossing of Schoharie Creek and the world's longest single-span covered bridge.

NEW JERSEY: Sunset in the Garden State

STRATEGICALLY WEDGED between the great cities of New York and Philadelphia, New Jersey has always been in the forefront of waterway, road and railway development. It is logical, therefore, that the state can claim a number of "firsts" in the chronology of covered bridges.

The Delaware River on the western boundary, draining nearly 3000 square miles of the state and thus of considerable geographic importance, was a covered bridge proving ground. Three great covered bridge builders—Timothy Palmer, Theodore Burr and Lewis Wernwag—tested their designs with wooden spans over the Delaware. Each of the three experimental designs was in some way novel, and all were toll bridges.

A private company wanted to span the river from Trenton to Morrisville, Pennsylvania. Theodore Burr was hired to build a 1008-foot bridge which cost $180,000, a staggering sum for those days. Forty years later the Trenton bridge was still considered "one of the finest specimens of wooden bridge architecture in the world."

A bridge such as this had never before been attempted. Huge laminated arches were fashioned by the York State builder out of four-inch planks which were 35 to 50 feet long. The floor was hung from the arches by means of iron rods, making the bridge technically "an arch span with suspended roadway"—actually two carriage lanes and two sidewalks. Even the roof was unconventional. Rather than use a gable roof, Burr had the five spans covered with the roof peaks crossways above the tops of the arches. Built in 1805, this strange-looking structure was the second covered bridge to be built in America and the first one between two states.

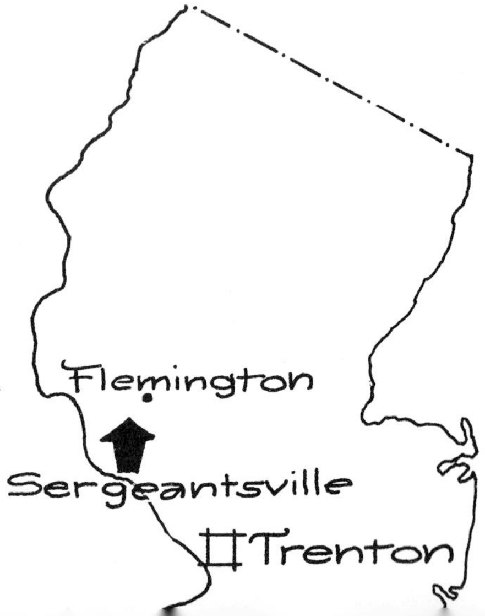

Timothy Palmer, builder of America's original covered bridge down at Philadelphia, directed the completion of another Delaware River span later the same year. It crossed from Phillipsburg to Easton, Pennsylvania, and was a massive structure with closely clapboarded sides and glass windows in sliding sashes. The Easton Bridge was composed of three spans set on gigantic piers in the Delaware, and cost but half the price of the Trenton bridge.

The third big bridge venture in New Jersey's western valley was constructed nine years later under the direction of Lewis Wernwag, German political refugee and bridge builder extraordinary. Wernwag had just conquered the Schuylkill River at Philadelphia's Upper Ferry with a single-span structure 340 feet long. At Lambertville, New Jersey, the German builder erected an entirely different type of truss over the Delaware. It had six arched spans, each 175 feet long; and for the first time the arches rested on top of the abutments and piers—instead of being notched into their faces as the arches in earlier bridges had done. Wernwag had an interest in an iron works at Phoenixville, Pennsylvania, so the bridge was a pioneer in the use of the metal with iron diagonal braces running between the timber posts. The designer patented the plan and called it New Hope, in honor of the town on the Pennsylvania side of the river.

Soon other villages along the Delaware followed this lead and sold shares to build their own toll bridges along the river. Eventually there were nineteen covered highway and

Flemington
Sergeantsville
Trenton

Burr used an unorthodox roof on America's first interstate covered bridge at Trenton

railroad bridges spanning the Delaware from the New York line to Trenton. The people of Hunterdon and Bucks Counties called them after the towns they served, rarely adding "bridge" to the name. "We'll cross by Yardley," they would say, or "We'll go by Frenchtown."

The history of the Delaware River is in part a story of terrible floods. January of 1841 was the first big one on record—with houses, barns, coal boats, broken log rafts and cakes of ice all slamming downstream in a grand rush. Not a bridge was left intact between Easton and Trenton.

George B. Fell of Lambertville happened to be on Centre Bridge when it went out and broke into two massive pieces. Fearing that the sides would soon collapse on him, George crawled out of the bridge and, with a plank for a gangway, reached a wide floating piece of roof nearby without wetting his feet. On this roof he was carried down the river by the swift current. The water was nearly up to the bottom of New Hope Bridge, but Fell lay flat on his roof piece and skimmed under it. Behind him a big chunk of Centre Bridge took out a New Hope arch with a rending crash. A span wrenched loose on the Jersey side, and a third span collapsed as a pier of Lewis Wernwag's pride crumbled.

Now the river behind Fell was a mass of bridge wreckage, ice chunks and roiling water. He steeled himself for a rough voyage downstream. At Taylorsville (now Washington's Crossing) seven miles farther on, some men tried to throw him a rope, but his raft lurched and he missed it. At Yardley, Fell banged into a bridge pier, lost most of his roof fragment and was drenched with icy water. No sooner had he passed than the whole Yardley bridge crumpled into the Delaware. George began to gather as much floating debris as he could reach, and with a makeshift paddle finally guided his tiny raft into some still water above Trenton. Fell had come seventeen miles unscathed down the angry Delaware. The broken bridges went charging past him out in midstream and passed harmlessly under Theodore Burr's high-piered Trenton Bridge. The weary voyager returned to Lambertville, where the villagers welcomed him with cheers and shot off an old cannon to celebrate his safe return.

The Trenton oddity in bridges and Timothy Palmer's masterpiece at Phillipsburg-Easton never did succumb to high water. The Camden & Amboy Railroad adapted half of the Trenton span for railroad use, and it became the second interstate covered railroad bridge in America. Burr had certainly never envisioned iron horses crossing the Delaware on his bridges, but they continued to do so until

These views of Timothy Palmer's Phillipsburg Bridge show the span's great strength and fine detail.

The last big Delaware River bridge joined Columbia and Portland, Penna., stood from 1869 to 1955.

the old wooden fabric was replaced by an iron structure in 1876. Palmer's bridge at Easton had to be replaced twenty years later —not from old age, but simply because it was desired to run trolley cars across the Delaware to Phillipsburg, and the roadways of the bridge were just not big enough.

The big flood of 1903 did away with another batch of Delaware River covered bridges. Once more seven out of the eight between Easton and Trenton were either partly or completely destroyed. New Hope broke loose and sailed off, her lamps still burning when the wreckage was found stranded in a field four miles downstream. Only Centre Bridge at Stockton was left intact, since it had been rebuilt a prudent six feet higher after the flood of 1841. Centre Bridge survived another twenty years, to be struck by lightning and burned during a summer thunderstorm.

A new flood, caused by hurricane-driven rains concentrated in the famous Water Gap region, took out the last big Delaware River covered bridge in August, 1955. This was the crossing between Columbia, New Jersey, and Portland, Pennsylvania. It had been built in 1869 and was the longest (760 feet) covered bridge in the United States at the time. The bridge resisted the raging waters for longer than anyone would have dreamed, but at last all but the span on the Jersey side went drifting off down the river it had crossed so faithfully for eighty-six years.

There were never a great many covered bridges wholly within New Jersey. Among the handful were the crossings at places like Little Falls and New Brunswick over the Passaic and Raritan Rivers. Two of the three crossings at Three Bridges, New Jersey, were once covered structures. On the tidal streams of south Jersey were a few more like those at Salem, Port Elizabeth and Dividing Creek, while the tributaries of the Delaware River on the western border were spanned by others. At Bridgeboro there once stood an old covered bridge dating from 1838, with gracefully arched portals and pedimented columns. Up above were painted eagles with outstretched wings and the national motto on a ribbon grasped in their claws.

At South Branch a former covered bridge across the Raritan River achieved belated fame by being featured in the late Robert Ripley's "Believe It or Not" cartoon feature in 1941. Ripley stated that the bridge was "over 200 years old." Covered bridge historians immediately took exception to this statement in view of the fact that the South Branch Bridge was built on the Town lattice plan, which wasn't invented until 1820!

No account of New Jersey's covered bridges is complete without mention of three half-forgotten spans on pioneer railroad lines in the northeastern part of the state. On the Paterson & Hudson River Railroad were the nation's first railroad drawbridges, finished in 1833. These were Long truss structures of the colonel's own design, with the erection superintended by Ireland's gift to American covered bridge building, Thomas Hassard. The draw over the Hackensack River was devised so that a whole section of the bridge swung aside, while the Passaic River bridge had a counterbalanced draw section.

That the winds really blow across the Jersey meadows is evidenced by Hassard's account of an accident that occurred shortly after the Hackensack Bridge was put into use. The railroad still had no locomotives and the cars were horse-drawn. On a December day the draw platform of the bridge was blown aside by a frigid blast, and the car thrown down the bank. No one was injured. The driver set out to fetch another car, but as he and the horse emerged from the protection of the bridge both he and the horse were blown into the ditch. Later, the bridge was the scene of one of America's first railroad fatalities, occurring when a passenger thrust his head out of the window as his train passed through the bridge. The railroad was not held to blame. Mr. Hassard simply stated that the poor fellow had been "imprudent" to get himself killed that way.

The tidy span at Sergeantsville is the only existing descendant of New Jersey's wooden giants.

A bit to the south was the bridge across the Raritan River, built to bring the New Jersey Railroad into New Brunswick. Completed in 1837, this was the first known double-decker railroad and highway bridge. It carried the railroad track on its flat tin roof and the highway in its dark interior. Another product of Long-Hassard talent, New Brunswick Bridge, carried main line traffic on the Pennsylvania Railroad for more than forty years.

Another railroad engineer, Simeon S. Post of Jersey City, patented in 1863 a combination wood-and-iron bridge truss which had only a spotty success. A few of this design were built in scattered areas in the Northeast, both on railroads and highways, and occasionally they were weatherboarded and roofed. For a brief period this bridge was a favorite in northern New Jersey, where the Watson Manufacturing Company of Paterson, assisted by the inventor's son Andrew, built Post trusses at dozens of sites.

Only a single covered bridge stands in New Jersey today. This is the small span over Wickecheoke Creek between Sergeantsville and Rosemont in the west-central portion of the state. This Hunterdon County landmark stands at the foot of a hill at an open spot in the beautiful wooded valley, its board-and-batten finish painted a neat white. Its construction is a local adaptation of the queenpost truss which outlines a framework of heavy wooden cross braces and iron rods. The date of erection is thought to have been around 1866. The proud giants that spanned the Delaware, the pioneer railroad jobs and the big and little covered bridges that crossed the rocky streams and marshy tidal rivers have all gone, leaving the Sergeantsville span as sole survivor of the covered bridge in New Jersey.

VI

Smoke Under the Eaves:

THE RAILROAD BRIDGES

THE LOCAL freight slows as it rounds the the bend and the rumble of the diesel engine is muffled as the locomotive disappears inside the covered bridge. Then comes a squealing of flanged wheels, a creak of old timber and the stifled puff of exhaust. Emerging, the engine gathers speed and the short train is soon lost to view in the windings of the wooded valley. Quiet returns to the gaunt covered wooden bridge.

In only four spots in the United States can you witness a sight like this—and three of them are in the Northeast. One by one the railroad spans have passed from the scene, doomed by fire, flood and obsolescence. Far sooner than their highway counterparts they are joining the ranks of the cigar store Indian, the Mississippi Show Boat, the crystal set—and the steam locomotives they once carried.

This condition has come about gradually. An English traveler said in 1841: "The timber bridges of America are justly celebrated for their magnitude and strength. By their means the railways of America have spread widely and extended rapidly." Without low-cost wooden bridges thousands of miles of rail lines in the United States would have been delayed or perhaps never built. Wood was the material of early American railroading, and wood it had to be for all the trial-and-error building that went on throughout the first forty years of railroad expansion.

Some of the first railroads used covered wooden bridges that were already built, sharing the passageway with vehicular traffic. Such was the case at Schenectady, New York,

This fine Long truss bridge carried the little Tonnewanta Railroad over the Erie Canal in Rochester.

where for six years (1833–39) the rails of the little Saratoga & Schenectady Railroad crossed the Mohawk River on the Northeast's unique wooden suspension bridge. This ingenious structure had been built twenty-four years earlier by Theodore Burr, a man who never saw a train but whose master craftsmanship permitted the iron horse to use the span in safety. Some years later another Burr-built covered bridge across the Delaware River at Trenton, New Jersey, was laid with tracks to support the Camden & Amboy Railroad. Covered bridges of the Burr type built exclusively for railroad use were a feature of the chain of little lines that became the New York Central. He was not responsible for a poorly adapted later version of his truss, with flat arches, which served unsuccessfully for a few years at a number of crossings along the lakes to the sea route of the Erie Railroad.

Henry R. Campbell, a Pennsylvanian noted for early locomotive design and experimentation, introduced Burr arch railroad bridges into New England and northern New York in the 1840's. The directors of the Vermont Central Railroad brought him up from Philadelphia as chief engineer to build their road. Campbell sized up the Vermont terrain and soon had his crews busy erecting arch bridges across the Winooski, the Lamoille, the Connecticut and all the lesser rivers in between. Nearby railroads, like the Vermont & Canada, the Northern (N.H.) and the Northern (N.Y.) clamored for, and received, his assistance in designing their bridges. For most of his jobs Campbell used what was virtually a trademark: he had the spans sided in a manner that showed the great sweep of the arches.

One bridge he planned became a storm center. In 1852 the Boston, Concord & Montreal Railroad was completing its trackage through New Hampshire. At Woodsville it wished to cross the Connecticut River to form a junction with the Connecticut & Passumpsic River Railroad operating over on the far bank in Wells River, Vermont. To get things rolling the BC&M made an agreement with an old toll bridge company to build a brand-new bridge at the proposed crossing that could be used by both the highway and the steam

A hassle between rival railroads held up construction on the Wells River–Woodsville combination bridge. Trains skimmed along on the nearly flat roof, but horses faced cranky turns at either portal.

cars. They bought land on the Wells River side and commenced to put up an abutment. But the Connecticut & Passumpsic people regarded the New Hampshire outfit as an enemy invader and wanted to build their own bridge to connect with a newly-built and promising line, the White Mountain Railroad running up to Littleton.

Then began the so-called bridge war between the rival railroads. BC&M men would dump tons of dirt into the fill for their Vermont foundation, only to have a small army of Passumpsic boys cart it away in the small hours of the night. This state of affairs kept up for a week, until the Irish laborers wore their tempers thin. Clubs and shillelaghs were being freely used when the first constable was summoned from his bed in Newbury. He arrived just in time to prevent wholesale slaughter by reading the riot act to the angry mob. The disturbance ended only when the competing railroads finally arbitrated and reached a mutual agreement for joint use of the bridge.

When it was completed the Wells River-Woodsville combination bridge had a single span of 231 feet. Trains had clear going on the single track laid across the roof, which was nearly flat. But poor Dobbin, enjoined to keep to a walk despite an ominous rumble overhead, had to turn sharply to avoid collision with a wall of stone at either end of the bridge. For fifty years users popped in and out of the "rat-holes-in-the-wainscot" bridge entrances, and stopped to pay toll besides.

Two more of these double-decker bridges carrying both a railroad and a highway spanned the Connecticut farther down at Northfield and at Montague City, Massachusetts. To the east at Newburyport a third one of this type carried the Eastern Railroad across the wide Merrimack River.

The patent design of Colonel S. H. Long was the initial favorite on New England lines, and railroad bridges used this plan on all the roads reaching out, wheel-spoke fashion, from Boston. The New Jersey railroads followed suit. In New York State the Town plan was preferred and lines like the Rensselaer & Saratoga, the Schenectady & Troy and the Harlem Railroad had large spans of the lattice type.

A lattice bridge was a deciding factor in the passing of the short-lived Catskill & Canajoharie Railroad, which ran twenty-two miles from Catskill up to Cooksburg, New York. This line maintained what was called the High Rock Bridge spanning a deep wooded ravine through which flowed Catskill Creek. On March 4, 1840, High Rock Bridge suddenly fell apart and a train of cars went tumbling into the stream. One man, Jehiel Tyler of East Durham, was killed, perhaps the first fatality in a collapse of an American railroad bridge. The C&C, beset by lawsuits, never resumed operations.

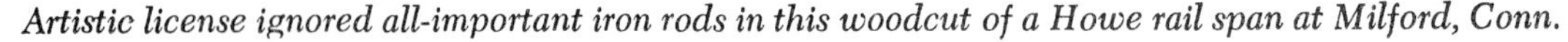

Artistic license ignored all-important iron rods in this woodcut of a Howe rail span at Milford, Conn.

The engineer thought the drawspan was closed when he came through the big covered bridge at Troy, N.Y., and the locomotive Jay Gould *of the Rensselaer & Saratoga Railroad plunged into the Hudson River.*

Within a very few years after it was invented in 1840 the patented Howe truss became the leading railroad bridge of the era and was built on virtually all the lines of the Northeast. The companies formed in Springfield, Massachusetts, by William Howe and his associates, contracted to build most of them.

The Howe companies would put up bridges while the railroad was actually being constructed. It was their custom to send gangs of men into the woods along the right of way to select good trees, fell them and hew them square with broadaxe and adze. The timbers would be piled to season awhile, and then were worked by an auger-and-chisel gang to ready them for assembly. Last of all, a third crew brought out the iron parts—rods, washers, nuts and bearing blocks—and erected the spans on stonework which masons had already prepared at end of track. Twenty-four hours would often see a 100-foot Howe truss bridge in place and ready to be roofed and sided after it was in use. At first most of the iron was imported from Scotland. Then foundries were established in Springfield to supply the demand from all the companies hard at work from Maine to Jersey and west to Buffalo building the patent truss.

The covered railroad bridge developed a standard design and began to look entirely different from its highway brethren. The trusses were high for ample clearance, sometimes as high as twenty-three feet. The bridges appeared narrower, seldom had any decoration, and those that were painted soon became sooty. The covered railroad span was strictly utilitarian and hardly a thing of beauty. There were exceptions—like the bridge at Bellows Falls, Vermont, which had stone portals set into its entrances, and William Howe's big job at Springfield, Massachusetts, that sported towering Egyptian columns at the ends. Romance could never flourish in a covered railroad bridge: the longer the bridge the more smoke found its way into the cars to start the poor passengers coughing and choking. Only the best bridges had smoke vents in the tops of their roofs.

In the decades following the Civil War covered bridges stood staunch on many a railroad in the Northeast. A survey of all the truss bridges in the State of New York—published in 1891—showed a surprising total of 418 bridges out of 2500 to be of wooden construction. As late as 1900 there were more than

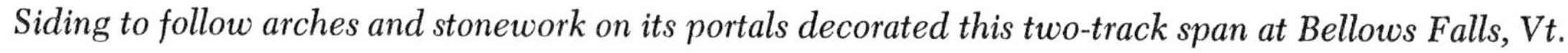

Siding to follow arches and stonework on its portals decorated this two-track span at Bellows Falls, Vt.

Double-decker at New Brunswick, N.J.

100 covered wooden bridges on the main and branch lines of the Boston & Maine's New Hampshire Division. On this railroad near Lebanon, New Hampshire, trains popped in and out of fourteen covered bridges across the Mascoma River in a space of six miles!

The B&M was unusual in that it sponsored a big revival in building covered railroad bridges. This boom was sparked by two engineers, David Haselton of the Concord Railroad and Willis Pratt of the Eastern—lines that the B&M absorbed. The structures were tightly enclosed, with heavy lattice trusses and with laminated arches sometimes built in or added later in order to carry heavier traffic. Fifty years ago a 120-foot iron bridge cost $5300 to build, a Howe truss $5000 and a Town lattice only $3500. Small wonder that the B&M made Town's its standard bridges and proudly declared that, if they were cared for properly, they would last "indefinitely" on branch lines. The company was still building them of local lumber to replace older lattice and Howe truss bridges as recently as 1906.

B&M-built covered bridges were strong, as is evinced by stories of accidents that they survived. A span over the mouth of the Contoocook River in New Hampshire was endangered by a heavy ice jam brought on by a sudden thaw. Loaded gravel cars were pushed onto the bridge and left there to weigh down the structure and press it more firmly against the abutments, a practice continued to this day in the face of such a hazard. Although the ice was several feet above the rail level the bridge was saved from serious damage by this weight and by its extra arches seated deep in the abutments.

One B&M bridge was the scene of a rear-end collision between trains in which an entire freight car was pushed upward and outward through the roof. Another bridge, near Woodsville, New Hampshire, was deroofed when a wrecking crew strange to the line forgot to lower the boom of their Big Hook. In both cases only new roofs and a few cross braces were necessary to repair the bridges.

The covered railroad bridges of New England have gradually disappeared, three or four in each decade. The former Rutland Railroad used an old covered Howe truss at East Shoreham, Vermont, until 1951, when it scrapped its Addison Branch to buy new diesels. (The bridge, minus rails, still stands.) The New Haven Railroad's only covered bridge, over the Blackstone River at Woonsocket, Rhode Island, succumbed to Hurricane Diane in 1955.

Fire was always a great enemy of any bridge with wooden sheathing, and the railroad spans were particularly vulnerable. Firemen

Hallmark of rail bridges was the massive strength shown in former StJ&LC span near Hardwick, Vt.

had orders to close their ashpans while crossing bridges. Further protection was afforded by an inside coat of whitewash, and water barrels, buckets and a ladder at each bridge. Many annual railroad reports have an item for "bridge-watching," payments to watchmen stationed at every major crossing to inspect the structure for sparks or hot ashes after each woodburner had trundled across. Boston & Maine covered bridges at Kelley's Falls (Manchester), Mast Yard, Bennington and Goffstown in New Hampshire were destroyed by spectacular flames in recent years, but the fires were not of railroad origin.

One by one the old-timers pass on, until today there are only eight covered railroad bridges on three common carriers and one tourist attraction line in New Hampshire and Vermont. First sight of them never fails to startle the casual observer: they are so immense compared with a highway bridge and so notable for their appearance of massive strength.

The last remaining covered bridge on the B&M is a 219-footer at Hillsboro, New Hampshire.

Until 1954 the B&M maintained three more covered bridges on its line between Concord and Claremont Junction, New Hampshire. Then the bridges became part of a new short line, the Claremont & Concord Railway. One bridge stands right in the center of the village of Contoocook, and the other two span

A Boston & Maine local sped over the covered bridge near Chandler, N.H., on its way to Concord.

Former covered railroad bridge at Bennington, N.H.

Interior siding which yet allowed trusses to be inspected, a full-length smoke vent and the odd keystone portals combined to make the former rail bridge at Haverhill, Mass., unusual.

the Sugar River near Chandler. With abandonment of the Claremont & Concord, the rails are gone and the bridges stand rotting and unused.

Only one of the covered bridges once maintained by the St. Johnsbury & Lamoille County Railroad in northern Vermont still carries trains: a ninety-footer east of Wolcott which has been reinforced with hidden steel supports. This is the nation's shortest existing covered railroad bridge. For contrast, at Swanton is StJ&LC's abandoned giant, which remains the longest covered railroad bridge in all America. Its three spans stretch for 369 feet across the Mississquoi River. With dieselization of the StJ&LC the odd cupola-like smoke vents on the roof of the bridge have been removed, and the Swanton pigeons have had to seek new quarters.

The covered bridge across the Winooski River east of Montpelier, Vermont, was the last Howe Truss used by a scheduled railroad carrier in the Northeast. This rarity among covered bridges was built in 1904 to serve the Barre Branch of the Montpelier & Wells River line. Later it was part of the Barre & Chelsea and still later carried locomotives of the Montpelier & Barre Railroad. When the last-named road took over the parallel trackage of a Central Vermont branch, the old Howe truss was purchased by Clark's Trading Post of Lincoln, New Hampshire. Moved in 1964 to a new site there spanning the Pemigewasset River, it now carries the cars of the shortline Clark Railroad.

The bridge formerly spanning the Winooski River east of Montpelier, Vermont.

VII

Today and Tomorrow

THE ONE THING responsible for the greatest boom in covered bridges also caused them to be discarded. This was an iron rod, a portion of William Howe's patented truss of 1840, and it prompted builders to erect hundreds of highway and railroad bridges of the Howe design from Maine to California.

The rod's function in the beginning was a simple one: it replaced the wooden uprights in Long and multiple kingpost trusses and—with its washers and bolts and turnbuckles—it allowed a span's members to be tightened when the bridge tired under the impact of traffic. Soon, however, the rods began to be used as actual supports for roadways and metal angle blocks appeared in the joints of the timbers. Iron was taking over.

It is not just sentimentality that makes people deplore the neglect and destruction of covered bridges in favor of ones built entirely of iron and erected in such profusion between 1870 and 1900. To see their point we must look for a moment at the relative merits of wood and of the iron used in those days. Wood has always been one of the finest building materials. Strong in itself, it is in a certain degree elastic, and therefore it can be compressed from both ends; and when it is compressed it becomes even stronger. But the quality that increases its strength under compression is responsible for its weakness when it is joined together: the weak parts of an all-wood truss are the joints. In iron trusses, on the other hand, the strongest parts are the joints, even though a beam fashioned from the brittle iron used early in bridge construction might not have possessed the potential stoutness of a great wooden timber. Comparison of the strengths of covered bridges and their immediate replacements in iron would make an interesting study. How many wooden spans would still be usable if they had received the scraping, painting and repairs given to iron bridges? It would help, too, to know how many covered bridges were replaced because their dimensions or their right-angle location on roads made them awkward crossings for automobiles. Disasters, like the floods of 1927, took out a great number of wooden bridges—around 200 in Vermont were carried away by high water—but this figure does not mean that iron spans were stronger. After all, with the higher cost of labor and materials and with road plans improved to drain traffic into main thoroughfares, iron bridges have never been erected with the same profusion that covered bridges were thrown across streams at every small crossing. Naturally our old covered bridges could not be expected to serve modern highways in and out of congested areas, but it is impossible to estimate how many wooden bridges off the beaten paths in small communities were doomed simply because they were "old fashioned" in the eyes of town planners.

Be all this as it may, covered bridges were threatened with eventual extinction. Then, around forty years ago, people in the Northeast grew alarmed over the rate at which wooden bridges were being destroyed, and they began spending time and energy to keep them from being pulled down. During the next two decades the rate of deliberate replacement of covered bridges with modern structures averaged only six a year in New England and New York. At the same time, however, Ohio initiated a wholesale program of building new bridges and thus reduced the total number of covered bridges from 600 to less than 300. One North Carolina county, which had fifty roofed spans in 1937, now has only two. With the example of the Northeast before them, however, other regions may be expected to cut down the number of replacements in the years to come.

There are five basic types of covered bridge fans. Because of the popularity of photography, foremost are those who like to take pictures of the bridges. The geographers, or "back roads addicts," find the *search* for covered bridges a prime enjoyment. Then there are the true historians who endeavor to track down the stories behind each bridge by talking to people in the neighborhood and delving through musty town records. Those with an interest in engineering find fascination in the study of the varied bridge truss designs. Finally, there are armchair hobbyists. Circumstance or necessity keeps them close to home, but often they accumulate a vast store of data to go along with a fine collection of pictures. Perhaps they have actually seen only a mere dozen covered bridges, but they are able to experience a secondhand thrill from the accounts and pictures of others. Each of these types possesses some or all of the others' approaches to the hobby and together they promote a well-rounded interest in saving these bridges.

With so many kinds of enthusiasts cropping up, it was inevitable that a society would be formed to devote itself to covered bridges. The first one in the Northeast was founded in 1949 by an elevator salesman, the late George B. Pease, and a concert musician, Leo Litwin. It was incorporated as The National Society for the Preservation of Covered Bridges, or NSPCB for short, and its headquarters are in Beverly, Massachusetts.

Originally a chapter of NSPCB, but now an independent organization, is The Connecticut River Valley Covered Bridge Society in Greenfield, Massachusetts. Sparkplug of this group is Mrs. Orrin H. Lincoln, an indefatigable worker in the cause of collecting and preserving covered bridges. There is also a New York State Covered Bridge Society, headquartered in Rochester.

Old-fashioned stereoscopic views, like this one of Brattleboro, Vt., whet collectors' detective ability.

These societies hold regular meetings, print fine quarterly publications devoted to covered bridges past and present and promote tours of existing bridges all over the Northeast. Even more important, they have been a real factor in the preservation of several covered bridges. Sometimes it is simply a letter to the selectmen, calling attention to an excessive number of loose boards on a bridge. In other cases a small cash donation will serve as a start for a town to appropriate funds to put its bridge or bridges in good shape again. Both societies point out, however, that the *will* to save a covered bridge must come from the structure's own neighbors: covered spans are seldom preserved as a result of pressures from outside the communities where they stand.

Hundreds of visitors saluted Old Blenheim on its 100th birthday, and in doing so they sparked a wider interest in old spans and paid tribute to the integrity of covered bridges everywhere.

Today all the publicity departments of the various states in the Northeast are acutely aware of the value their covered bridges possess as tourist attractions. Vermont, New Hampshire and Massachusetts locate all of theirs on detailed maps. In addition the Green Mountain state's Historic Sites Commission is working hard and well to prevent fine examples from being torn down and to mark others with plaques describing their history. Maine and New York publish illustrated leaflets about their covered bridges, while Connecticut and New Jersey often make note of roofed spans in press releases and advertising. Local chambers of commerce and garden clubs have been instrumental in putting up signs to direct sightseers to a nearby covered bridge or two. Tourism is a business bringing in millions of dollars to the region, so the attraction of covered bridges is certain to be noted and the structures will be allowed to stand wherever it is possible to let them do so. We have already seen efforts on the part of New Hampshire and Massachusetts to replace their covered bridges in kind. On May 15, 1957, Governor Edmund S. Muskie of Maine signed a bill authorizing a fund of $50,000 in state aid to help any city or town that desires to maintain an existing covered bridge. This legislation should ensure the continued existence of the nine spans left in Maine, and it is a fine precedent for other states to follow.

The covered bridge hobby is not like collecting antiques, where one can cart home his treasures in the family car. Instead one must be content with photographs, paintings and drawings. Fans can debate for hours over which is the best camera to use for covered bridge pictures. The late E. H. Royce, dean of Vermont covered bridge photographers, said the choice depends upon each person's requirements, but he recommended a 35mm. camera as practical for either color or black-and-white film. A flash gun should be used in photographing interiors, he adds, in order to reduce the extreme contrast between the darkness inside and the bright light at the entrances. Exterior photographs should be made from different angles: the ideal one is a three-quarter position which shows an entrance portal, one side of the bridge and the stream it spans; road level, stream level and a higher elevation are all good, he felt, where trees and shrubbery permit an uncluttered view. Head-on pictures are a dime a dozen, however, and should be taken only to show some special feature of the bridge. Because the angle of the light is important for interesting black-and-white pictures, he suggested early morning or late afternoon as the best times, because then the shadows bring out patterns and details.

Almost a separate sub-hobby is that of searching out old photographs, woodcuts, engravings and lithographs of covered bridges that have long ceased to stand. The hunt for such mementos can become as exciting as detective work. Among the oldest of these are woodcuts and etchings of the earliest known covered bridges. Photographs, taken from old tintypes, start before the Civil War, as do the double stereoscopic views; and these, too, often show covered bridges that stood for only a few short years. Occasionally a long-sought picture of a certain bridge turns up in a place far removed from the site of the span itself and becomes a cherished item in the researcher's collection.

After 1900, picture postcards held sway, and on them are recorded the likenesses of hundreds of covered bridges which once stood on highways and byways and whose appearance might otherwise be unknown today. Many of these cards are to be found in old scrapbooks and albums and as loose assortments in secondhand bookstores. One of the strong arguments in favor of collecting pictures of these spans is that, unlike accumulating coins, stamps or birds' eggs, one does not require originals in order to have a fine collection. A photographic copy of an old card or stereo is just as good as, and sometimes turns out better than, the faded print from which it was made. Another specialty is that of making a collection of modern postcards showing covered bridges, now that reproductions of Kodachrome photographs are generally available.

It is good to watch covered bridges coming into their own in the Northeast, to see states and towns concerned over their landmarks and private capital restoring or duplicating fine examples. For an interest in old covered bridges can be a completely satisfying thing, since they embrace so many kindred subjects. Anyone who goes in for them much is bound to become better acquainted with American history and geography. Delving more deeply, he can learn painlessly something of engineering, carpentry, masonry and the properties of wood. With knowledge comes a heartfelt respect for the genius of an earlier day, and soon this regard warms to delight in each span for its own sake and fosters the hope that, perhaps just around the next bend, will be standing a covered bridge with all its nostalgic charm.

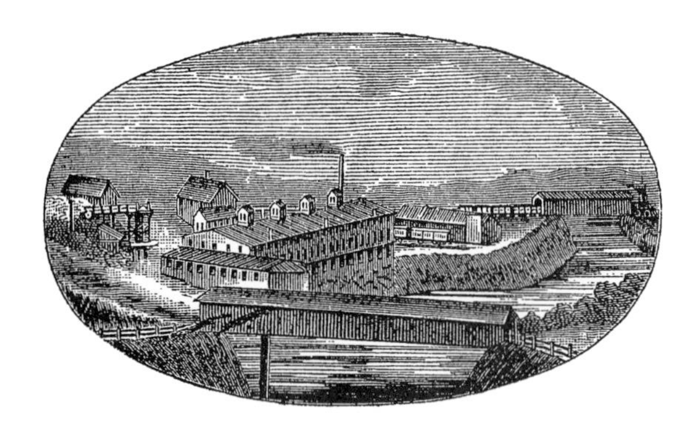

APPENDIX I. TABULATION OF EXISTING COVERED BRIDGES

This list stresses age and authentic old-time construction. It omits foot bridges, fakes, bridges over dry land and structures that span nothing. A scattering of "modern" bridges on public highways has been included. A few private spans of recent construction may well prove worthy of being added. R.S.A., 1 Jan. 1981

Location	Name	County	Stream	No. Spans, Length in ft	Date Built	Builder	Type of Construction	Remarks
MAINE								
E. of Littleton	Watson settlement	Aroostook	Meduxnekeag Stream	2–160	1911		Howe	
1½ m. NW of S. Windham	Babb's	Cumberland	Prescumscot River	Unknown				Modern replacement of 1864 bridge
South Andover	Lovejoy	Oxford	Ellis River	1– 80	1867		Paddleford	
NW of East Fryeburg	Hemlock	Oxford	Old Saco River	1–116	1857		Paddleford	
S. of Wilsons Mills	Bennett	Oxford	Magalloway River	1–100	1901		Paddleford	
NW of North Bethel	Sunday River	Oxford	Sunday River	1–100	1872	Hiram York	Paddleford	
Porter	Porter-Parsonfield	Oxford-York	Ossipee River	2–160	1858		Paddleford	
Robyville	Robyville	Penobscot	Kenduskeag Stream	1– 76	1876		Howe	Fully Shingled
Bangor	Morse	Penobscot	Kenduskeag Stream	2–212	1884		Howe	Moved & Preserved 1965
E. of Sangerville	Lowe's	Piscataquis	Piscataquis River	1–130	1857		Long	
NEW HAMPSHIRE								
E. of Bartlett		Carroll	Saco River	1–183			Paddleford	Preserved
Conway	Conway	Carroll	Saco River	2–240	1890	Charles & Frank Broughton	Paddleford	Preserved
Albany		Carroll	Swift River	1–136	1858	Amzi Russell & Leander S. Morton	Paddleford	
Conway		Carroll	Swift River	1–144	1870	Jacob Berry & Jacob Berry, Jr.	Paddleford	
Jackson	Jackson	Carroll	Ellis River	1–138	1876	Charles Broughton	Paddleford	Sidewalk
West Ossipee	Whittier	Carroll	Bear Camp River	1–144		Jacob Berry	Paddleford	
W of North Sandwich	Durgin	Carroll	Cold River	1–110	1869	Jacob Berry	Paddleford	
W of Swanzey	Sawyer's Crossing	Cheshire	Ashuelot River	2–159	1859		Town	
West Swanzey		Cheshire	Ashuelot River	2–159	1832	Zadoc Taft	Town	Sidewalk
Westport	Slate	Cheshire	Ashuelot River	1–122	1862		Town	
S. of Westport	Coombs	Cheshire	Ashuelot River	1–118	1887		Town	
Ashuelot		Cheshire	Ashuelot River	2–160	1864		Town	Sidewalk
E. of Swanzey	Carleton	Cheshire	So. Branch of Ashuelot River	1– 60			Queenpost	
W. of Pittsburg		Coos	Connecticut River	1– 91			Paddleford	Closed
E. of Pittsburg	Happy Corner	Coos	Perry Stream	1– 86			Paddleford	
E. of Pittsburg		Coos	Perry Stream	1– 57			Queenpost	
Stark		Coos	Upper Ammonoosuc River	2–151			Paddleford	Sidewalk

LOCATION	NAME	COUNTY	STREAM	NO. SPANS, LENGTH IN FT	DATE BUILT	BUILDER	TYPE OF CONSTRUCTION	REMARKS
Groveton		Coos	Upper Ammonoosuc River	1–136	1852	Charles Richardson & Son	Paddleford	Sidewalk
Lancaster	Mechanic Street	Coos	Israel River	1–108	1862		Paddleford	
Columbia Bridge	Columbia	Coos	Connecticut River	1–148	1912	Charles Babbitt	Howe	
South Lancaster	Mount Orne	Coos	Connecticut River	2–285	1911	Babbitt Bros. & Berlin Construction Co.	Howe	
Bath		Grafton	Ammonoosuc River	4–356	1832		Burr	
Woodsville		Grafton	Ammonoosuc River	2–278	1829		Town	Sidewalk
Swiftwater		Grafton	Wild Ammonoosuc Riv.	1–174	1849		Paddleford	
Haverhill	Bedell's	Grafton	Connecticut River	2–436	1866		Burr	Closed
Flume Reservation	Flume	Grafton	Pemigewasset River	1– 35	1886		Paddleford	
Blair Station	Blair	Grafton	Pemigewasset River	2–273	1869		Long	
W. of Plymouth	Smith	Grafton	Baker River	1–160	1844	Charles Richardson	Long	
E. of Beebe River	Bump	Grafton	Beebe River	1– 66	1877		Queenpost	
N. of Lyme	Edgell	Grafton	Clay Brook	1–154	1885	J.C. & W.G. Piper	Town	
N. of West Campton	Turkey Jim	Grafton	West Branch Brook	1– 60	1874		Queenpost	
Hancock–Greenfield	County	Hillsboro	Contoocook River	1– 88	1937	Hagen-Tibideau Const. Co.	Teco	
Henniker	New England College	Merrimack	Contoocook River	1–130	1972	Milton S. Graton	Town	
West Hopkinton	Rowell's	Merrimack	Contoocook River	1–167	1852		Long	
Bradford	Bement	Merrimack	Warner River	1– 63	1854		Long	
Waterloo		Merrimack	Warner River	1– 76	1857		Town	
Warner	Dalton	Merrimack	Warner River	1– 75			Multiple Kingpost	
W. of Potter Place	Cilleyville	Merrimack	Blackwater River	1– 49	1887		Town	Closed
Andover	Keniston	Merrimack	Blackwater River	1– 71	1882	A.R. Hamilton	Town	
NW of Dover	County Farm	Strafford	Cocheco River	1–112	1879		Howe	
Cornish "City"	Blacksmith Shop	Sullivan	Mill Brook	1– 90	1882	James F. Tasker	Multiple Kingpost	
Cornish "Mills"	Dingleton	Sullivan	Mill Brook	1– 81	1882	James F. Tasker	Multiple Kingpost	
W. of Meriden	Mill	Sullivan	Blood's Brook	1– 80	1880	James F. Tasker	Multiple Kingpost	
S. of Langdon	Prentiss	Sullivan	Great Brook	1– 36	1874	Sanford Granger	Town	Preserved
NE of Alstead		Sullivan	Cold River	1– 78	1869	Sanford Granger	Town	
North Newport	Corbin	Sullivan	Croydon Branch Sugar River	1–105			Town	
Cornish	Windsor	Sullivan	Connecticut River	2–460	1866	James F. Tasker & Bela J. Fletcher	Town	

VERMONT

W. of Salisbury Sta.		Addison	Otter Creek	1–136	1865		Town	
NW of Middlebury	Pulp Mill	Addison	Otter Creek	3–179	c.1820		Burr	Double-Barrel, Orig. 1 span
NE of Middlebury	Halpin	Addison	Muddy Branch of New Haven River	1– 56	1840		Town	41 feet above stream
W. of North Ferrisburg	Spade	Addison	Gully	1– 85	1850	Justin Miller	Town	(moved to site)
East Shoreham		Addison	Lemon Fair River	1–108	1897	Rutland RR	Howe	Ex-railroad Bridge
NW of Bennington	Silk	Bennington	Walloomsac River	1– 88	1840		Town	
NW of Bennington	Papermill Village	Bennington	Walloomsac River	1–125	1889	Charles F. Sears	Town	
S. of North Bennington	Henry	Bennington	Walloomsac River	1–121	1840		Town	
W. of West Arlington	Arlington Green	Bennington	Batten Kill	1– 66	1852		Town	
N. of East Arlington	Chiselville	Bennington	Roaring Branch of Batten Kill	1–117	1870	Daniel Oatman	Town	
W. of Lyndon	Chamberlain	Caledonia	So. Wheelock Branch of Passumpsic River	1– 66			Queenpost	Closed
Lyndon	Schoolhouse	Caledonia	So. Wheelock Branch of Passumpsic River	1– 42	1872		Queenpost	
N. of Lyndon Center	Bradley	Caledonia	Miller Run	1– 56			Queenpost	
NE of Lyndonville	Lang	Caledonia	East Branch of Passumpsic River	1– 57			Queenpost	
South Danville	Greenbank Hollow	Caledonia	Joes Brook	1– 50	1886		Queenpost	
S. of Prindle Corner		Chittenden	Lewis Creek	1– 58			Burr	
S. of East Charlotte	Quinlan	Chittenden	Lewis Creek	1– 87	1849		Burr	
NW of Charlotte		Chittenden	Home Creek	1– 39			Laminated Arch	
Westford		Chittenden	Browns River	1– 97			Burr	Closed
Shelburne		Chittenden	Burr Pond	1–168	1845	Farewell Wetherby	Burr	Double-Barrel (moved here 1951; Blt. orig. at Cambridge over Lamoille River)
Montgomery	Comstock	Franklin	Trout River	1– 80	1883	Jewett Bros.	Town	
NW of Montgomery	Harnois	Franklin	Trout River	1– 89	1863	Jewett Bros.	Town	
NW of Montgomery	Hopkins	Franklin	Trout River	1– 80	1875	Jewett Bros.	Town	
Montgomery	Fuller	Franklin	Black Falls Brook	1– 55	1890	Jewett Bros.	Town	
S. of Montgomery Ctr.	Hectorville	Franklin	S. Branch, Trout River	1– 54	1883	Jewett Bros.	Town	
S. of Montgomery Ctr.	Hutchins	Franklin	S. Branch, Trout River	1– 54	1883	Jewett Bros.	Town	
SW of Montgomery	West Hill	Franklin	West Hill Brook	1– 40	1883	Jewett Bros.	Town	Closed (Private)
Fairfax		Franklin	Mill Brook	1– 57			Town	
East Fairfield		Franklin	Black Creek	1– 68	1865		Queenpost	
E. of Belvidere	Morgan	Lamoille	N. Br. of Lamoille River	1– 65	1895	Lewis Robinson	Queenpost	
N. of Belvidere	Lumber Mill	Lamoille	N. Br. of Lamoille River	1– 72	1895	Lewis Robinson	Queenpost	
N. of Waterville		Lamoille	N. Br. of Lamoille River	1– 62	1895		Queenpost	
N. of Waterville		Lamoille	N. Br. of Lamoille River	1– 65			Queenpost	
Waterville		Lamoille	N. Br. of Lamoille River	1– 72			Queenpost	
East Johnson		Lamoille	Gihon River	1– 60			Queenpost	
Johnson	Power Plant	Lamoille	Gihon River	1– 65			Queenpost	
SE of Johnson	Waterman	Lamoille	Waterman Brook	1– 70			Queenpost	
Cambridge Junction	Poland	Lamoille	Lamoille River	1–153	1887	George W. Holmes	Burr	
S. of Jeffersonville	Scott	Lamoille	Brewster River	1– 80			Burr	
Cambridge	Little Cambridge	Lamoille	Seymour River	1– 60	1897	George W. Holmes	Burr	Moved here 1950 (Private)

LOCATION	NAME	COUNTY	STREAM	NO. SPANS, LENGTH IN FT	DATE BUILT	BUILDER	TYPE OF CONSTRUCTION	REMARKS
E. of Sterling	Sterling	Lamoille	Sterling Brook	1– 66			Queenpost	
SE of Stowe	Stowe Hollow	Lamoille	Gold Brook	1– 49			Howe	
Thetford Center		Orange	Ompompanoosuc River	1– 80			Haupt	With Auxiliary Arch
Union Village		Orange	Ompompanoosuc River	1–100	1867		Multiple Kingpost	
S. of Chelsea	Moxsley	Orange	First Br. of White River	1– 55	1887	Arthur Adams	Queenpost	
N. of North Tunbridge	Flint	Orange	First Br. of White River	1– 50	1845		Multiple Kingpost	
N. of North Tunbridge	Bates	Orange	First Br. of White River	1– 55	1902		Multiple Kingpost	
Tunbridge	Hayward & Noble	Orange	First Br. of White River	1– 60	1883		Multiple Kingpost	
S. of Tunbridge	Cilley	Orange	First Br. of White River	1– 65	1883		Multiple Kingpost	
N. of South Tunbridge	Howe	Orange	First Br. of White River	1– 60	1879		Multiple Kingpost	
S. of East Randolph		Orange	Second Branch of White River	1– 45	1904		Multiple Kingpost	Half-size Truss
S. of East Randolph		Orange	Second Branch of White River	1– 50	1904		Multiple Kingpost	Half-size Truss
N. of East Bethel		Orange	Second Branch of White River	1– 50	1904		Multiple Kingpost	
SW of Coventry		Orleans	Black River	1– 87	1881	John D. Colton	Paddleford	
S. of Irasburg		Orleans	Lord Creek	1– 50	1881	John D. Colton	Paddleford	Private
S. of North Troy	Upper	Orleans	Missisquoi River	1– 91			Town	With single pins
N. of Proctor	Gorham	Rutland	Otter Creek	1–114	1841	Abraham Owen & N.M. Powers	Town	
							Town	
Pittsford Station	Depot	Rutland	Otter Creek	1–121			Town	
Florence Station	Hammond	Rutland	Otter Creek	1–139	1842	Asa Nourse	Town	Preserved
S. of Brandon	Dean	Rutland	Otter Creek	1–136	1865		Town	
SW of Brandon	Sanderson	Rutland	Otter Creek	1–132			Town	
East Clarendon	Kingsley	Rutland	Mill River	1–120	1870	T.K. Horton	Town	
E. of North Clarendon	Brown	Rutland	Cold River	1–100	1880	Nicholas M. Powers	Town	
S. of Pittsford	Cooley	Rutland	Furnace Brook	1– 60	1849	Nicholas M. Powers	Town	80 foot Ridgepole
Warren		Washington	Mad River	1– 37	1880	Walter Bagley	Queenpost	
Waitsfield	Old Arch	Washington	Mad River	1–120	1833		Burr	Sidewalk
N. of Waitsfield Common		Washington	Pine Brook	1– 40	1872		Kingpost	
S. of Northfield Falls	Slaughter House	Washington	Dog River	1– 55			Queenpost	
Northfield Falls		Washington	Dog River	1–100	1872		Town	
W. of Northfield Falls		Washington	Cox Brook	1– 45			Queenpost	
W. of Northfield Falls		Washington	Cox Brook	1– 54			Queenpost	
SW of Northfield		Washington	Rocky Brook	1– 39	1899		Kingpost	
E. of Plainfield	Orton Farm	Washington	Winooski River	1– 50	1890	Herman F. Townsend	Queenpost	Private
W. of Plainfield	Coburn	Washington	Winooski River	1– 50	1851		Queenpost	
Green River		Windham	Green River	1–104	1870		Town	
Townshend State Forest	Scott	Windham	West River	3–276	1870	Harrison Chamberlain	Town	Plus: 2 spans Kingpost
N. of West Dummerston		Windham	West River	2–280	1872	Caleb B. Lamson	Town	
W. of Williamsville		Windham	Marlboro Branch	1–120	1870		Town	

W. of Brattleboro	Creamery	Windham	Whetstone Brook	1– 80	1879	A.H. Wright	Town	Sidewalk
E. of Saxtons River	Hall	Windham	Saxtons River	1–117	1982	Milton S. Graton	Town	Replacement of 1867 Br.
Grafton	Kidder	Windham	S. Branch of Saxtons River	1– 67	1870		Queenpost	
Bartonsville		Windham	Williams River	1–151	1870	Sanford Granger	Town	
S. of Bartonsville		Windham	Williams River	1– 87	1868	Sanford Granger	Town	
W. of Rockingham	Victorian Village	Windham	Br. of Williams River	1– 42	1872	Harrison Chamberlain	Queenpost	Moved here 1967; Orig. at West Townshend
W. of Amsden	Upper Falls	Windsor	Black River	1–121			Town	
North Springfield	Baltimore	Windsor	Br. of Great Brook	1– 42	1870	Granville Leland	Town	Moved to site 1970
W. of Brownsville	Best's	Windsor	Mill Brook	1– 36			Laminated Arch	
W. of Brownsville		Windsor	Mill Brook	1– 44			Laminated Arch	
W. of West Woodstock	Lincoln	Windsor	Ottauquechee River	1–134	1877	R.W. & B.H. Pinney	Pratt	Adaption with Arch
Woodstock	Middle	Windsor	Ottauquechee River	1–125	1969	Milton S. Graton	Town	
Taftsville		Windsor	Ottauquechee River	2–189	1836	Solomon Emmons	Queenpost-Arch	
North Hartland	Willard	Windsor	Ottauquechee River	1–123			Town	
E. of Hartland	Martin's Mill	Windsor	Lull's Brook	1–119	1881	James F. Tasker	Town	
MASSACHUSETTS								
Sheffield	Upper	Berkshire	Housatonic River	1– 93	1835		Town	
Sheffield		Berkshire	Housatonic River	1–135	1953		Teco	
Gilbertville		Hampshire-Worcester	Ware River	1–135	1886		Town	
Conway	Burkeville	Franklin	South River	1–107	1873		Multiple Kingpost	Iron Rod Verticals
N. of Greenfield	Pumping Station	Franklin	Green River	1– 94	1972		Teco	Replacement of 1870 Br.
S. of Lyonsville	Arthur Smith	Franklin	North River	1–112	1870		Burr	Re-erected here 1896
N. of Charlemont	Bissell	Franklin	Mill Brook	1– 92	1951		Teco	
East Pepperell		Middlesex	Nashua River	1– 76	1963		Teco	
Old Sturbridge Village		Worcester	Arm of Quinebaug Riv.	1– 65	1870		Town	Orig. "Taft" Bridge in Dummerston, Vt.; to Sturbridge 1952; moved to present site 1956
CONNECTICUT								
West Cornwall		Litchfield	Housatonic River	2–242	1841		Town	
Bulls Bridge		Litchfield	Housatonic River	1–109	1842		Town	
Colchester–Easthampton	Comstock's	Middlesex–New London	Salmon River	1– 80	1873		Howe	Truss Variation
NEW YORK								
Downsville		Delaware	E. Br. Delaware River	1–174	1854	Robert Murray	Long	

Location	Name	County	Stream	No. Spans, Length in Ft	Date Built	Builder	Type of Construction	Remarks
East Delhi	Fitch's	Delaware	W. Br. Delaware River	1–100	1885	David L. Wright	Town	Orig. blt. in Delhi in 1870 by James Frazier & James Warren
Hamden		Delaware	W. Br. Delaware River	2–128	1859	Robert Murray	Long	
SW of Margaretville	Tuscarora Club	Delaware	Mill Brook	1– 24	1870		Kingpost	Orig. blt. at Dunraven; moved here 1935 (Private)
S. of Cook's Falls	Methol	Delaware	Trout Brook	1– 24	1877	Anson Jenkins & Augustus Neidig	Town	44 ft. span cut to 24 ft.; moved here 1954
Jay		Essex	E. Br. Ausable River	4–150	1857		Howe	Originally single-span
Salisbury Center		Herkimer	Spruce Creek	1– 49	1875	Alvah Hopson	Burr	
S. of East Springfield	Hyde Hall	Otsego	Shadow Brook	1– 40	1830		Burr	In Glimmerglass State Park
Buskirk		Rensselaer–Washington	Hoosic River	1–165	1880		Howe	
Edinburg		Saratoga	Beecher Creek	1– 29	1879	Arad Copeland	Queenpost	Private
North Blenheim	Blenheim	Schoharie	Schoharie Creek	1–210	1855	Nicholas M. Powers	Long	Double-barrel, unique center arch
Hall	Hall's Mills	Sullivan	Neversink River	1–119	1912	David Benton & John Knight	Town	Closed
Beaver Kill	Camp Site	Sullivan	Beaver Kill	1– 98	1865	John Davidson	Town	
W. of Willowemoc	Bendo	Sullivan	Willowemoc Creek	1– 43	1860	John Davidson	Town	Cut in half & moved here 1913
W. of Livingston Manor	Van Tran Flat	Sullivan	Willowemoc Creek	1– 98	1860	John Davidson	Town	
Newfield		Tompkins	W. Br. Cayuga Inlet	1– 80	1853	Samuel Ham & Sons	Town	
Rifton	Perrine's	Ulster	Wall Kill	1–138	1844		Burr	
S. of Seager		Ulster	Dry Brook	1– 32	1907	Jerome Moot	Kingpost	
S. of Seager		Ulster	Dry Brook	1– 42	1907	Jerome Moot	Kingpost	
N. of Seager	Forge	Ulster	Dry Brook	1– 27	1907	Jerome Moot	Kingpost	
SW of Margaretville	Grants Mills	Ulster	Mill Brook	1– 66	1902	Edgar Marks & Wesley Alton	Town	Closed
E. of Olive Bridge		Ulster	Old Channel Esopus Creek	1– 60		Frank Mead	Town	Orig. at Turnwood; moved here 1930 (Private)
East Salem	Eagleville	Washington	Batten Kill	1– 88	1858		Town	
Shushan		Washington	Batten Kill	2–160	1858	Stevens Bros.	Town	Preserved on concrete pedestal
S. of Salem	Rexleigh	Washington	Batten Kill	1–100	1874		Howe	
NEW JERSEY								
W. of Sergeantsville		Hunter Don	Wickecheoke Creek	1– 84	1866		Queenpost	

RAILROAD BRIDGES

Location	Railroad	Stream	No. Spans, Length in Ft	Date Built	Type	Remarks
Hillsboro, N.H.	Boston & Maine	Contoocook River	2–219	1903	Town-Pratt	Sidewalk
Contoocook, N.H.	Claremont & Concord	Contoocook River	1–157	1889	Town-Pratt	Unused
E. of Chandler, N.H.	Claremont & Concord	Sugar River	2–228	1906	Town-Pratt	Unused
W. of Chandler, N.H.	Claremont & Concord	Sugar River	1–122	1905	Town-Pratt	Added laminated arch; unused
Lincoln, N.H.	Clark's Trading Post	Pemigewasset River	1–100	1904	Howe	Orig. east of Montpelier, Vt.; Moved here 1964
E. of Wolcott, Vt.	St. Johnsbury & Lamoille County	Lamoille River	1– 90	1908	Town-Pratt	
Swanton, Vt.	St. Johnsbury & Lamoille County	Missisquoi River	3–369	1898	Town-Pratt	Unused

APPENDIX II. DISTRIBUTION BY TYPE

TYPES OF CONSTRUCTION IN THE NORTHEAST	KINGPOST & VARIETIES	QUEENPOST & VARIETIES	MULTIPLE KINGPOST & VAR.	BURR-KINGPOST ARCH	TOWN LATTICE	LONG (ALL WOOD)	HOWE (WOOD AND IRON)	PADDLEFORD	SIMPLE LAMINATED ARCH	PRATT ADAPTATION (WOOD & IRON)	HAUPT	"TECO"	DOUBLE WEB TOWN-PRATT RR LATTICE	COMBINATION TRUSSES	TOTALS
MAINE						1	3	5							9
NEW HAMPSHIRE *		4	4	1	15	4	3	14				1			46
VERMONT †	2	23	9	9	38		2	2	3	1	1			2	92
MASSACHUSETTS			1	1	3							4			9
CONNECTICUT					2		1								3
NEW YORK	4	1		3	11	2	3							1	25
NEW JERSEY		1													1
RAILROAD BRIDGES							1						6		7
TOTALS	6	29	14	14	69	7	13	21	3	1	1	5	6	3	192

* Plus 5 railroad bridges (4 Town-Pratt Lattice. 1 Howe)

† Plus 2 railroad bridges (Town-Pratt Lattice)

APPENDIX III – ROSTER OF BUILDERS

BOODY, AZURIAH—*Born* Stanstead, Que., April 21, 1815; *died* New York, N.Y., November 18, 1885.

BOOMER, GEORGE B.—*Born* Sutton, Mass., July 26, 1832; *died* Vicksburg, Miss., May 22, 1863.

BOOMER, LUCIUS B.—*Born* Douglas, Mass., July 4, 1826; *died* New York, N.Y., March 6, 1881.

BRAINERD, EZRA—*Born* East Haddam, Conn., May 11, 1769; *died* Holley, N.Y., Nov. 15, 1833.

BRIGGS, ALBERT D.—*Born* Brattleboro, Vt., January 25, 1820; *died* Springfield, Mass., February 20, 1881.

BROUGHTON, FRANK—*Born* Conway, N.H., 1864; *died* Conway, N.H., November 1943.

BURR, THEODORE—*Born* Torringford, Conn., August 16, 1771; *died* Middletown, Penna., November 21, 1822.

CHILDS, ENOCH—*Born* Henniker, N.H., August 8, 1808; *died* Henniker, N.H., September 8, 1881.

CHILDS, HORACE—*Born* Henniker, N.H., August 10, 1807; *died* Henniker, N.H., June 8, 1900.

CHILDS, WARREN—*Born* Henniker, N.H., October 12, 1811; *died* Henniker, N.H., April 10, 1888.

DAMON, ISAAC—*Born* Weymouth, Mass., July 16, 1792; *died* Northampton, Mass., December 4, 1862.

DAVIDSON, JOHN—*Born* New York, N.Y., 1815; *died* Shin Creek (Lew Beach), N.Y., 1875.

FIELD, REUBEN—*Born* Cranston, R.I., Nov. 22, 1772; *died* Troy, N.Y., 1842.

FLETCHER, BELA J.—*Born* January 16, 1811; *died* Claremont, N.H., July 26, 1877.

GRANGER, SANFORD—*Born* Chesterfield, N.H., 1796; *died* Bellows Falls, Vt., 1882.

HARRIS, DANIEL—*Born* Providence, R.I., February 6, 1818; *died* Springfield, Mass., July 11, 1879.

HASELTON, DAVID—*Born* Dorchester, N.H., 1832.

HASSARD, THOMAS—*Born* Ireland.

HAUPT, HERMAN—*Born* Philadelphia, Penna., March 26, 1817; *died* Jersey City, N.J., Dec. 14, 1905.

HAWKINS, RICHARD F.,—*Born* March 9, 1837; *died* Springfield, Mass., March 5, 1913.

HOWE, WILLIAM—*Born* Spencer, Mass., May 12, 1803; *died* Springfield, Mass., September 19, 1852.

LONG, MOSES—*Born* Hopkinton, N.H.; *died* Rochester, N.Y., March 1857.

LONG, STEPHEN H.—*Born* Hopkinton, N.H., December 30, 1784; *died* Alton, Illinois, September 4, 1864.

MURRAY, ROBERT—*Born* Eskdale, Scotland, November 22, 1814; *died* New Brunswick, N.J., December 19, 1898.

PADDLEFORD, PETER—*Born* Enfield, N.H., September 14, 1785; *died* Littleton, N.H., October 18, 1859.

PALLADIO, ANDREA—*Born* Vicenza, Italy, 1518; *died* Vicenza, 1580.

PALMER, TIMOTHY—*Born* Rowley, Mass., August 22, 1751; *died* Newburyport, Mass., December 19, 1821.

PARKER, GEORGE A.—*Born* Concord, N.H., May 8, 1822; *died* Lancaster, Mass., April 20, 1887.

POST, SIMEON S.—*Born* Lebanon, N.H., August 29, 1805; *died* Jersey City, N.J., June 28, 1872.

POWERS, NICHOLAS M.—*Born* Pittsford, Vt., August 30, 1817; *died* Clarendon, Vt., January 17, 1897.

PRATT, WILLIS T.—*Born* Boston, Mass., July 4, 1812, *died* Boston, Mass., July 10, 1875.

SPOFFORD, MOODY—*Born* Rowley, Mass., June 24, 1744; *died* Georgetown, Mass., Dec. 23, 1828.

STEWART, DANIEL—*Born* Hadley, N.Y., April 15, 1791; *died* Hadley, Aug. 5, 1881.

STONE, AMASA, JR.—*Born* Charlton, Mass., April 27, 1818; *died* Cleveland, Ohio, May 11, 1883.

STONE, ANDROS B.—*Born* Charlton, Mass., June 18, 1824; *died* New York, N.Y., December 15, 1896.

STONE, DANIEL—*Born* Charlton, Mass., May 28, 1810; *died* Philadelphia, Penna., November 26, 1863.

STONE, JOSEPH—*Born* Charlton, Mass., February 16, 1808; *died* Springfield, Mass., October 8, 1873.

TASKER, JAMES F.—*Born* Cornish, N.H., September 15, 1826; *died* Claremont, N.H., July 18, 1903.

TOWN, ITHIEL—*Born* Thompson, Conn., October 3, 1784; *died* New Haven, Conn., June 13, 1844.

WALCOTT, JONATHAN—*Born* Windham, Conn., April 23, 1776; *died* Windham Centre, Conn., November 13, 1819.

WERNWAG, LEWIS—*Born* Riedlingen, Germany, December 4, 1769; *died* Harpers Ferry, Va., August 12, 1843.

WHIPPLE, SQUIRE—*Born* Hardwick, Mass., September 16, 1804; *died* Albany, N.Y., March 15, 1888.

WHISTLER, GEORGE W.—*Born* Fort Wayne, Ind., May 19, 1800; *died* St. Petersburg, Russia, April 7, 1849.

WOODS, DUTTON—*Born* Henniker, N.H., 1809; *died* 1884.

Selected Bibliography

Babcock, John B., 3rd., *The Boston Society of Civil Engineers and Its Founder Members.* Boston, 1936. (Reprinted from Boston Society of Civil Engineers, Journal, July, 1936.)

Bell, William E., *Carpentry Made Easy.* Philadelphia, 1857.

Brangwyn, Frank, and Sparrow, Walter S., *A Book of Bridges.* London and New York, 1915.

Congdon, Herbert Wheaton, and Royce, Edmund Homer, *The Covered Bridge.* Brattleboro, Vt., 1941.

Cooper, Theodore, *American Railroad Bridges.* New York, [1889].

Covered Bridge Topics, edited by Richard S. Allen, Eugene R. Bock and Leo Litwin. National Society for the Preservation of Covered Bridges, April 1943–

Dana, Richard T., *The Bridge at Windsor, Vt., and its Economic Implications.* New York, 1926.

Edwards, Llewellyn N., *The Evolution of Early American Bridges.* Paper #15, Maine University, Maine Technology Experiment Station, Orono, Maine, 1934. Reprinted from an address before the Newcomen Society of North America.

Essex Institute, *Historical Collections; Timothy Palmer, Bridge Builder of the Eighteenth Century.* Salem, Mass., April 1947.

Fletcher, Robert, and Snow, J. P., *A History of the Development of Wooden Bridges.* Paper # 1864, American Society of Civil Engineers, New York, 1934.

Haupt, Herman, *General Theory of Bridge Construction.* New York, 1851.

Hosking, William, *Bridges, in Theory, Practice and Architecture.* London, 1839.

Jakeman, Adelbert M., *Old Covered Bridges.* Brattleboro, Vt., [1935].

Kirby, Richard S., and Laurson, Philip G., *The Early Years of Modern Civil Engineering.* New Haven and London, 1932.

Long, Col. S. H., *Description of Col. Long's Bridges.* Concord, N.H., 1836.

———, *Description of the Jackson Bridge.* Baltimore, 1830.

Mahan, D. H., *An Elementary Course of Civil Engineering.* New York, 1869.

Merriman, Mansfield, and Jacoby, Henry S., *A Text-Book on Roofs and Bridges.* New York, 1898.

Morley, S. Griswold, *The Covered Bridges of California.* Berkeley, 1938.

[Morse, Victor], *The Story of Covered Bridges in Windham County, Vermont.* Brattleboro, Vt., 1937.

New York State Board of Railroad Commissioners, *Annual Report.* Albany, 1856.

———, *Report of the Board of Railroad Commissioners of the State of New York on Strains on Railroad Bridges of the State.* Albany, 1891.

Pain, William, *The Carpenter's Pocket Directory.* Philadelphia, 1797.

Peale, Charles W., *An Essay on Building Wooden Bridges.* Philadelphia, 1797.

Pope, Thomas, *A Treatise on Bridge Architecture.* New York, 1811.

Reynolds, James, *Andrea Palladio.* New York, [1948].

Sloane, Eric, *American Barns and Covered Bridges.* New York, [1954].*

———, *Our Vanishing Landscape.* New York, [1955].*

Spargo, John, *Covered Wooden Bridges of Bennington and Vicinity.* Bennington, Vt., 1953.

Steinman, David B., and Watson, Sara Ruth, *Bridges and Their Builders.* New York, [1941].

Town, Ithiel, *A Description of Ithiel Town's Improvement in the Principle, Construction and Practical Execution of Bridges.* New York, 1839.

Trautwine, John C., *The Civil Engineer's Pocket-Book.* New York and London, 1891.

Tyrrell, Henry G., *History of Bridge Engineering.* Chicago, 1911.

Waddell, J. A. L., *Bridge Engineering.* New York, 1916.

Wagemann, Clara E., *Covered Bridges of New England.* Rutland, Vt., 1952.

Watson, Wilbur J., *Bridge Architecture.* New York, [1927].

Wells, Rosalie, *Covered Bridges in America.* New York, 1931.

Whipple, Squire, *An Elementary and Practical Treatise on Bridge Building.* New York, 1872.

———, *A Work on Bridge Building.* Utica, N.Y., 1847.

White, W. Edward, *Covered Bridges of New Hampshire.* Plymouth, N.H., 1942.

Whitney, Charles S., *Bridges: A Study in Their Art, Science and Evolution.* New York, 1929.

Wood, De Volson, *Treatise on the Theory of the Construction of Bridges and Roofs.* New York, 1873.

World Guide to Covered Bridges, edited by Richard Donovan. National Society for the Preservation of Covered Bridges, Inc. Beverly, Mass., 1980.

*For these and other titles on the same and related subjects, log on to www.doverpublications.com.

Glossary

ABUTMENT—The shore foundation upon which a bridge rests, usually built of stone but sometimes of bedrock, wood, iron or concrete.

ARCH—A structural curved timber, or arrangement of timbers, to support a bridge, usually used in covered bridges together with a truss. Thus a *supplemental* or *auxiliary arch* is one which assists a truss; a *true arch* bridge is entirely dependent upon the arch for support.

BEARING BLOCK (or ANGLE BLOCK)—Triangular block of wood or iron at the junction of post, brace, counter-brace or arch, serving as a seat for the members.

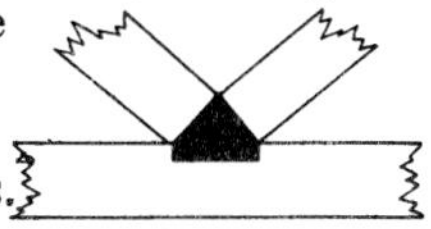

BENT—An arrangement of timbers resembling a sawhorse which is placed under a bridge at right angles to the stringers, sometimes used as a temporary scaffolding in building a covered bridge. Also to support light, open approaches, weak or damaged bridges, and sometimes as a substitute for abutments or piers.

BRACE—A diagonal timber in a truss which slants toward the mid-point of the bridge.

BUTTRESS—An outside timber or iron rod, one of whose ends joins an extended floor beam (q.v.) and whose upper end inclines inward and is attached to the truss to give lateral support. Sometimes called a *sway brace*.

CAMBER—A slight convexity, upward bowing or "hump" of the chords, built in to allow the bridge to be level after it settles.

CHORD—The top (*upper chord*) or bottom (*lower chord*) member or members of a bridge truss, usually formed by the stringers; may be a single piece or a series of long joined pieces.

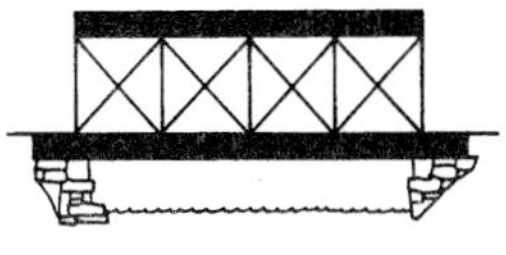

COMPRESSION MEMBER—A timber or other truss member which is subjected to squeeze. Often a diagonal member, such as a brace (q.v.) or counterbrace (q.v.).

CORBEL—In covered bridges, a solid piece of wood —mainly for decoration—which projects from the portal and assists in supporting the overhanging roof. Also, on a larger scale, a solid timber at the angle of an abutment (or pier) and lower chord, to lend extra support.

COUNTER-BRACE—A diagonal timber in a truss which slants away from the mid-point of the bridge (opposite from brace, q.v.).

DECK TRUSS—A type of bridge where the traffic, usually railroad, uses the roof on top of the truss as a roadbed; sometimes also carries traffic inside, between the trusses.

DOUBLE-BARRELED BRIDGE—Common designation for a covered bridge with two lanes; the divider can be a third truss or structural part of the bridge, or it can be a simple partition.

FACE OF ABUTMENT—The side of the abutment toward the center of the stream.

FALSEWORK—See SCAFFOLDING.

FLOOR BEAM—Transverse beam between bottom chords of trusses on which longitudinal joists are laid.

INFERIOR BRACE—Timber slanting upward from the face of the abutment to the underside of the bridge.

JOIST—Timbers laid longitudinally on the floor beams of a bridge and over which the floor planking is laid.

KEY—Piece, often a wedge, inserted in a joint such as a mortise and tenon to tighten the connection. Sometimes called a *fid*.

LAMINATED ARCH—A series of planks bolted together to form an arc; constructed in such a manner that the boards are staggered to give extra strength.

LATERAL BRACING—An arrangement of timbers between the two top chords or between the two bottom chords of bridge trusses to keep the trusses spaced apart correctly and to insure their strength. The arrangement may be very simple, or complex.

MORTISE, (n.)—Cavity made in wood to receive a tenon. (v.) To join or fasten securely by using a mortise and tenon.

NEEDLE BEAM—See FLOOR BEAM.

PANEL—Rectangular section of truss included between two vertical posts and the chords. A *panel system* is made up of three or more panels.

PATENTED TRUSS—Any one of the truss types for which United States patents have been granted, such as Burr, Town, Long, Howe, etc., trusses.

PIER—An intermediate foundation between abutments, built in the stream bed, for additional support for the bridge. May be made of stone, concrete, wood, etc.

PILE—Heavy timber, often a peeled log, sunk vertically into the stream bed to provide a foundation when the bottom is unreliable. Piling can be used as a base for abutments and piers, or the bridge can be built directly upon piling.

PORTAL—General term for the entrance or exit of a covered bridge; also used to refer to the boarded section of either end under the roof.

POST—Upright or vertical timber in a bridge truss; *center post* is the vertical timber in the center of a truss; *end post* is the vertical timber at either end of the truss.

RAFTER—One of a series of relatively narrow beams joined with its opposite number to form an inverted V to support the roof boards of a bridge.

ROD—Iron rod used as integral vertical member in some truss bridges to replace wooden posts between upper and lower chords. Bridge members could be tightened by adjusting nuts against washers on the ends of the rods. Their use marked the first step in transition from wooden bridges to bridges made entirely of iron.

SCAFFOLDING—Light, temporary wooden platforms built to assist in the erection of a bridge. Sometimes called *falsework*.

SCARF JOINT—A joint made by so notching the ends or sides of two timbers that they lock together and are then secured by bolts, etc., to form a virtually slip-proof joint.

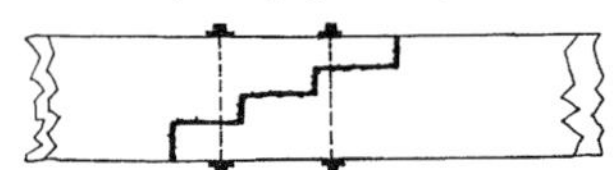

SECONDARY CHORD—Single or joined timbers lying between upper and lower chords and parallel to them, giving added strength to the truss.

SHELTER PANEL—The first panel at each end of both trusses of a panel-truss bridge, often boarded on the inside to protect the timbers from moisture blowing through the portals.

SHIP'S KNEE—A short timber bent at a right angle used inside a covered bridge between a truss and upper lateral bracing to increase rigidity. Similar to a corbel (q.v.) but heavier and not decorative. Sometimes called *knee brace*.

SIMPLE TRUSS—An elementary bridge truss, such as kingpost or queenpost; not so large or complex as the patented trusses.

SKEW-BACK—A jog or incline in the face of an abutment to receive the end of a chord or an arch.

SKEWED BRIDGE—A bridge built diagonally across a stream.

SPAN—The length of a bridge between abutments or piers. *Clear span* is the distance across the bridge, measuring from the face of one abutment to the face of the other. The length usually given is for the *truss span,* i.e., the length between one endpost of the truss and the other, regardless of how far the truss may overreach the actual abutment. Bridges of more than one span are called *multi-span bridges.*

SPLICE—A method of joining timbers, especially end-to-end, by means of a scarf or other joint, sometimes with keys or wedges inserted to give additional strength and stability to the joint. A *splice-clamp* is a metal or wooden clamp designed to hold two spliced timbers together.

STRINGER (or String-Piece)—A longitudinal member of a truss which may be made up of either one single timber, in comparatively short bridges, or a series of timbers spliced end-to-end in longer bridges. Most evident in the chords (q.v.) which often go by this name.

SUSPENSION ROD (or Hanger Rod or Suspender)—Iron rod usually found in arch bridges or in connection with auxiliary arches added to older bridges; attached from arch to floor beams to aid in supporting the roadway.

TENON—A tongue shaped at the end of a timber to fit into a mortise and so form a joint.

TENSION MEMBER—Any timber or rod of a truss which is subjected to pull, or stretch.

TREENAILS—Wooden pins which are driven into holes of slightly smaller diameter to pin members of lattice trusses together. (Pronounced "trunnels").

TRESTLE—A braced framework built up from the stream bed to support a bridge.

THROUGH TRUSS—A covered bridge in which traffic uses a roadway laid on the lower chords between the trusses. Most covered bridges are through trusses.

TRUSS—An arrangement of members, such as timbers, rods, etc., in a rigid form so united that they support each other plus whatever weight is put upon the whole. Covered bridge trusses, including arch trusses, employ a triangle or a series of combined triangles, since this is the form which cannot be forced out of shape by external pressure. *Truss* is also used to refer to just one side of a bridge.

TURNBUCKLE—A metal loop fashioned with a screw at one end and a swivel at the other, used in some covered bridge trusses to tighten iron rods and thus overcome sagging.

WEB—A truss design (such as Town lattice) in which timbers crisscross each other. A lattice truss, or a truss designed with overlapping panels, may be called a *web system.*

WEDGE—See Key.

WINDBRACING—Inside timbers extending from a point on a truss to the ridgepole to attach the roof more firmly to the sides of the bridge.

Index

ITALICIZED FIGURES INDICATE AN ILLUSTRATION

THIS BOOK, DESIGNED BY R. L. DOTHARD ASSOCIATES,

WAS SET IN CALEDONIA AND BULMER TYPES.